Praise for *The 33*

Previously titled Deep Down Dark

'An astonishing tale of survival' *Spectator*

'Tobar plunges the reader into this world of uncertainty with visceral, present-tense prose and careful pacing . . . Whether the story is completely new to you, or if you were one of the millions glued to the news reports and wondering, will they make it – physically, emotionally, spiritually – you'll be greatly rewarded to learn how they did.' *New York Times Book Review*

'An eloquent testament to the human spirit.'
Robert Crampton, *The Times*

'An account that brims with emotion and strength.' *USA Today*

'A gripping narrative, taut to the point of explosion . . . An electrifying, empathetic work of journalism that makes a four-year-old story feel fresh.' *Kirkus*

'His narrative cracks along at a suitably breathless pace . . . you're unlikely to find a more exciting account.' *Daily Mail*

'Weaving together the drama of the miners' harrowing ordeal below ground with the anguish of families and rescuers on the surface, Tobar delivers a masterful account of exile and human longing, of triumph in the face of all odds. Taut with suspense and moments of tenderness and replete with a cast of unforgettable characters, *The 33* ranks with the best of adventure literature.' *LA Times*

'Chiseled, brooding . . . As Tobar works his way through each miner's recovery, the TV headlines recede from our memory, and a more delicate series of portraits emerges.' *Washington Post*

ALSO BY HÉCTOR TOBAR

FICTION

The Barbarian Nurseries

The Tattooed Soldier

NONFICTION

Translation Nation

Héctor Tobar is the son of Guatemalan immigrants and a native of the city of Los Angeles. He is the former Buenos Aires and Mexico City Bureau Chief for the *LA Times* and shared a Pulitzer for the paper's coverage of the 1992 riots. He is currently an LA-based columnist for the paper. He is the author of the critically acclaimed novel, *The Barbarian Nurseries*.

www.hectortobar.com

@TobarWriter

THE 33

The Untold Stories of 33 Men Buried
in a Chilean Mine, and the Miracle
that Set them Free

Previously titled Deep Down Dark

HÉCTOR TOBAR

SCEPTRE

FOR THE PEOPLE OF CHILE

First published in Great Britain in 2014 by Sceptre
An imprint of Hodder & Stoughton
An Hachette UK company

First published in paperback in 2015
This film tie-in edition published in 2015

1

An excerpt from *The 33* originally appeared in *The New Yorker*.

Grateful acknowledgement is made for permission to reprint 'It's as if I'm
pushing through massive mountains' from *The Poetry of Rilke* by Rainer
Maria Rilke, translated and edited by Edward Snow. Translation copyright
© 2009 by Edward Snow. Reprinted by permission of North Point Press, a
division of Farrar, Straus and Giroux, LLC.
Image on page 2: composite picture of the 33 miners by AFP/Getty Images.
'Inside the mine' illustration by Abi Darker © *The New Yorker* Condé Nast

A CIP catalogue record for this title is available from the British Library

Paperback ISBN 978 1 473 63510 4
eBook ISBN 978 1 444 75543 5

Printed and bound by Clays Ltd, St Ives plc

Hodder & Stoughton policy is to use papers that are natural, renewable
and recyclable products and made from wood grown in sustainable forests.
The logging and manufacturing processes are expected to conform to the
environmental regulations of the country of origin.

Hodder & Stoughton Ltd
Carmelite House
50 Victoria Embankment
London EC4Y 0DZ

www.sceptrebooks.co.uk

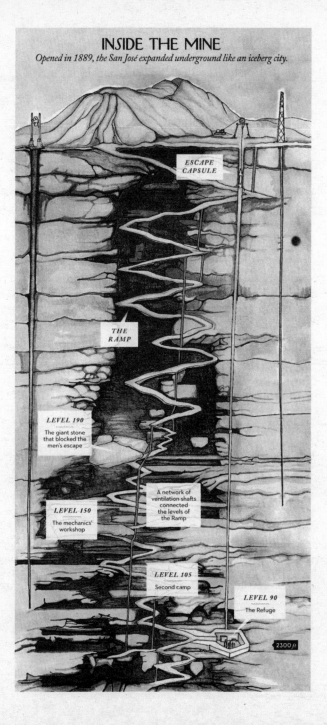

INSIDE THE MINE

Opened in 1889, the San José expanded underground like an iceberg city.

ESCAPE CAPSULE

THE RAMP

LEVEL 190

The giant stone that blocked the men's escape

A network of ventilation shafts connected the levels of the Ramp

LEVEL 150

The mechanics' workshop

LEVEL 105

Second camp

LEVEL 90

The Refuge

2300 ft

It's as if I'm pushing through massive mountains
through hard veins, like solitary ore;
and I'm so deep that I can see no end
and no distance: everything became nearness
and all the nearness turned to stone.

I'm still a novice in the realm of pain,—
so this enormous darkness makes me small;
But if it's *You*—steel yourself, break in:
that your whole hand will grip me
and my whole scream will seize you.

—Rainer Maria Rilke, *The Book of Hours*

CONTENTS

Prologue: Cities in the Desert   3

PART I: BENEATH THE MOUNTAIN OF THUNDER AND SORROW

1. A Company Man   19
2. The End of Everything   29
3. The Dinner Hour   43
4. "I'm Always Hungry"   55
5. Red Alert!   73
6. "We Have Sinned"   85
7. Blessed Among Women   103
8. A Flickering Flame   121
9. Cavern of Dreams   139

PART II: SEEING THE DEVIL

10. The Speed of Sound   163
11. Christmas   173
12. Astronauts   187
13. Absolute Leader   207
14. Cowboys   217
15. Saints, Statues, Satan   229
16. Independence Day   237
17. Rebirth   249

VIII CONTENTS

PART III: THE SOUTHERN CROSS

18. In a Better Country   267
19. The Tallest Tower   273
20. Underground   283
21. Under the Stars   295

Acknowledgments   307

DRAMATIS PERSONAE

THE MEN IN CHARGE, AND WHO TAKE CHARGE

Luis Urzúa, the shift supervisor
Florencio Avalos, foreman, Urzúa's assistant
Juan Carlos Aguilar, the supervisor of the mechanics
Mario "Perri" Sepúlveda, a front-loader operator

THE MECHANICS AND THE MEN FROM THE SOUTH

Raúl Bustos, lost his previous job thanks to a tsunami
José Henríquez, an Evangelical Christian known as "the Pastor"
Juan Illanes, a storyteller and labor-law authority
Edison Peña, electrician, athlete and Elvis fan
Richard Villarroel, a father-to-be who lost his own father as a child

THE OLDER MEN FROM THE NORTH

Yonni Barrios, fifty, the miner with two homes
Jorge Galleguillos, fifty-six, a whistleblower
Mario Gómez, sixty-three, missing fingers from a previous accident
Franklin Lobos, fifty-two, truck driver and a former soccer pro
José Ojeda, forty-seven, a widower who pens an important, pithy note
Omar Reygadas, fifty-six, white-haired, even tempered, spiritual
Esteban Rojas, forty-four, native of Copiapó and cousin to Pablo
Pablo Rojas, forty-five, veteran miner whose father died days earlier
Darío Segovia, forty-eight, working overtime for extra pay
Víctor Segovia, forty-eight, keeper of an underground diary

THE YOUNGER MEN FROM THE NORTH

Claudio Acuña, part of a "jumbo" crew, filmed videos underground
Osman Araya, an Evangelical Christian
Renán Avalos, younger brother to foreman Florencio
Samuel Avalos, nicknamed "CD" because he hawks pirated music
Carlos Barrios, becomes a leader of the men in The Refuge

Carlos Bugueño, fortification crew member
Pedro Cortez, will take charge of communications equipment
Daniel Herrera, fortification crew member
Carlos Mamani, a Bolivian immigrant
Jimmy Sánchez, eighteen and too young to legally work underground
Ariel Ticona, his girlfriend is about to give birth for the third time
Alex Vega, a homebody nicknamed "Pato," or Duck
Claudio Yáñez, becomes especially thin, listless, and physically weak
Víctor Zamora, grew up on the streets of Arica, near Peru

SOME FAMILY MEMBERS

María Segovia, sister of Darío Segovia
Jessica Chilla, girlfriend of Darío, mother to his daughter
Mónica, wife of Florencio Avalos
The Vega family:
 Jessica, wife of Alex Vega;
 José, Alex's father;
 Priscilla, his sister;
 Roberto Ramirez, Priscilla's boyfriend
Carmen Berríos, wife of Luis Urzúa
Elvira Valdivia, wife of Mario Sepúlveda
Francisco, their son
Scarlette, their daughter
Marta Salinas, Yonni Barrios's wife
Susana Valenzuela, Yonni's girlfriend

THE GOVERNMENT OFFICIALS

Sebastián Piñera, President of Chile
Laurence Golborne, the Minister of Mining
André Sougarret, veteran mine administrator, in charge of rescue
Cristián Barra, the president's fixer at the mine

THE DRILLERS

Eduardo Hurtado, supervisor of the Terraservice drill crew
Nelson Flores, a Terraservice driller
Jeff Hart, a veteran American driller
Pedro Gallo, a local entrepreneur and telephone expert

THE MINING COMPANY OFFICIALS

Carlos Pinilla, general manager of the San Esteban Mining Company
Alejandro Bohn and Marcelo Kemeny, owners of the San Esteban Mining
Company

THE 33

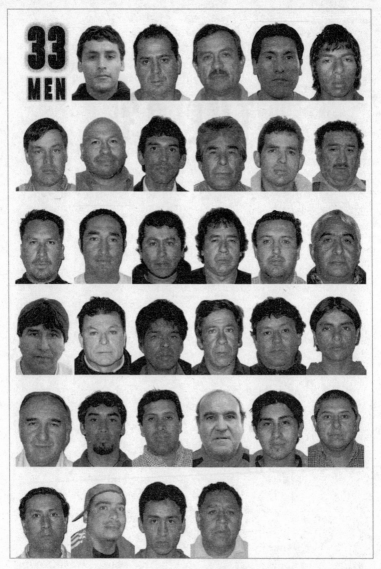

TOP ROW, FROM LEFT TO RIGHT: Florencio Avalos, Mario Sepúlveda, Juan Illanes, Carlos Mamani, Jimmy Sánchez; ROW 2: Osman Araya, José Ojeda, Claudio Yáñez, Mario Gómez, Alex Vega, Jorge Galleguillos; ROW 3: Edison Peña, Carlos Barrios, Víctor Zamora, Víctor Segovia, Daniel Herrera, Omar Reygadas; ROW 4: Esteban Rojas, Pablo Rojas, Darío Segovia, Yonni Barrios, Samuel Avalos, Carlos Bugueño; ROW 5: José Henríquez, Renán Avalos, Claudio Acuña, Franklin Lobos, Richard Villarroel, Juan Carlos Aguilar; ROW 6: Raúl Bustos, Pedro Cortez, Ariel Ticona, Luis Urzúa

PROLOGUE: CITIES IN THE DESERT

The San José Mine is located inside a round, rocky, and lifeless mountain in the Atacama Desert in Chile. Wind is slowly eroding the surface of the mountain into a fine, grayish orange powder that flows downhill and gathers in pools and dunes. The sky above the mine is azure and empty, allowing an unimpeded sun to bake the moisture from the soil. Only once every dozen years or so does a storm system worthy of the name sweep across the desert to drop rain on the San José property. The dust is then transformed into mud as thick as freshly poured concrete.

Few sightseers visit this corner of the Atacama, though the naturalist Charles Darwin did pass nearby, briefly, during his nineteenth-century journey around the world on a Royal Navy research vessel. The locals told him scientifically implausible stories that linked the rare rains to earthquakes. The vastness of Atacama and the absence of animal life surprised Darwin, and in his journal he described the desert as "a barrier far worse than the most turbulent ocean." Even today, birders who pass through this part of Chile note that there are few if any avian species to be found. In the deeper desert, the only conspicuous living presence is of mining men, and the occasional woman, riding in trucks and minibuses to the mountains where there is gold, copper, and iron to be extracted.

The wealth of minerals under the barren hills draws workers to the Atacama mines from the nearby city of Copiapó, and from many other distant corners of Chile. Juan Carlos Aguilar travels the farthest to reach the San José: more than a thousand miles. The shape of Chile on a map is that of a snake, and Aguilar's trek to work takes him along half the snake's body. His weekly commute begins in the temperate rain forests of southern Chile. At the mine he supervises a team of three men who repair front loaders and the squat, long-armed, insect-like vehicles known as "jumbos," during a seven-day shift that begins on Thursday morning. But his ride to work starts thirty-six hours earlier on Tuesday evening in the town of Los Lagos. No local job pays as well as his gig in the desert, so he settles his weary, middle-aged body into a Pullman bus, and watches as the shadows of beech forests, eucalyptus tree farms, and mountain rivers pass by his window. The weather matches his mood: The sky is overcast and raindrops beat on the windows, as they usually do when he sets out for work. The average precipitation in the region of Chile he calls home, at the 40th parallel of the Southern Hemisphere, is 102 inches a year.

One of the mechanics in the supervisor's three-man crew lives somewhat closer to the mine. Raúl Bustos leaves for work from the port city of Talcahuano, near the 37th parallel. A tsunami struck Talcahuano five months earlier, triggered by an 8.8 magnitude earthquake. The disaster killed more than five hundred people, left the city covered with pools of ocean water in which thousands of fish flopped about, and washed out the navy base where he used to work. Bustos is a punctilious father of two and devoted husband. He boards a bus heading northward, then travels along a flat landscape filled with greenhouses, tractors, and the fallow and cultivated fields of Chile's agricultural heartland. He passes through the town of Chillán, where another member of Aguilar's crew begins his own journey northward, and then through Talca, where a tall and devoutly Christian jumbo operator boards yet another bus. The men who work inside the San José Mine are divided into two shifts, A and B, each working seven days at a time, and all these men have been assigned to the A shift. The A shift, in turn, is divided into twelve-hour-long night and day shifts that keep the mine working around the clock, from 8:00 a.m. to 8:00 p.m. to the following 8:00 a.m.

The commuting members of the A shift soon enter Santiago, with its under-construction skyscrapers and elevated roads. For the southerners, it's early morning as they pull into Santiago, a booming Latin

American capital whose most distinguishing feature, the massive and soaring silhouette of the nearby Andes, is often lost in the haze of the city's notorious smog.

In the intercity bus terminals in the center of Santiago, not far from Chile's presidential palace, more men set out for the San José Mine. One is Mario Sepúlveda, a frenetic father of two who has a reputation among his coworkers for pushing the front loader he operates too hard (thus forcing the mechanics to repair it repeatedly), and for talking too loud and too much, and for being generally unpredictable. On Wednesday afternoon he sets out on the five-hundred-mile journey from Santiago to the San José Mine later than he should: There is a fair chance he won't make it to work on time. Mario's nickname at the San José is "Perri," which is short for "Perrito," the diminutive of the word *perro*, or "dog." Ask Mario why he's called Perri and he'll tell you it's because he loves dogs (he owns two rescued strays at home), and because "I have the heart of a dog." Like a dog, Mario is loyal, but if you try to hurt him, "I'll bite you." He and his wife, Elvira, have two children, the first conceived in an impassioned encounter "standing up, against a pole." Now they have a home on the fringes of Santiago where his prized possession is a big meat locker, and where Mario's favorite place to sit is the small, square-shaped table in his living room. He enjoys a hurried meal at the table with Elvira and their teenage daughter, Scarlette, and their young son, Francisco, before leaving for the north.

After departing central Santiago, and then passing through the working-class suburbs on the city's northern fringe, the various buses carrying the men of the A shift enter valleys filled with vineyards and fruit trees, the snow-covered Andes of August winter on their right. The climate is Mediterranean, but the vista loses more of its greenness with each passing hour and with each latitude they cross: 33°, 32°, 31° south. Soon they're entering the arid region called the Norte Chico, or Near North.

Mining men and other adventurers have traveled this route from the earliest days of Chilean history. The north is Chile's desert frontier, its Wild West. It's the place where the dictator Augusto Pinochet preferred to imprison his foes, gathering more than one thousand political dissidents in the living quarters of an abandoned saltpeter mine, where they passed the time studying astronomy underneath the pristine desert sky. Chile's union movement was born in the north, founded at the beginning of the twentieth century by nitrate miners who were later massacred in the city of Iquique, and in the democratic Chile of today much of the north still votes faithfully for the left. Pinochet also had the men and women he murdered buried in shallow graves in the desert, in the Norte Grande, or Far North, and their bones are still being discovered forty years later by relatives searching for the "disappeared."

When the men of the A shift reach the port city of Coquimbo, 250 miles from the San José Mine, they join the path that Charles Darwin followed in 1835 on the Chilean leg of his voyage. Chile was then a young country, merely twenty-five years old, and Darwin hopped off his ship, the HMS *Beagle*, to make observations about its geology and its flora and fauna, his small expedition riding overland with four horses and two mules. The road between Coquimbo and Copiapó passes through the oldest mining region in Chile, and the British naturalist met many miners on his slow trek along this same path.

In the town of Los Hornos, the men of the A shift catch a glimpse of the Pacific Ocean as Route 5, also known as the Pan-American Highway, passes along the beach. There's something cruel about getting this last look at the ocean and the horizon, with the late afternoon sun casting its warm rays upon the expansive surface of the water: For the next seven days, the men will be spending most of the daylight hours two thousand feet underground, in caverns just wide enough for their vehicles to squeeze through. In these working weeks of Southern Hemisphere winter they will see the sun only briefly, for a few minutes in the morning before their shift begins, and at lunchtime. Not far from the beach at Los Hornos, Darwin saw a hill that was being systematically mined, "drilled with holes, like a great ants' nest." He learned that the local miners sometimes earned great sums and then, "like sailors with prize-money," they found ways to "squander" their bonanza. The miners Darwin met drank and spent excessively, and in a few days returned "penniless" to their miserable jobs, there to work "harder than beasts of burden."

The men of the A shift do not expect to be penniless anytime soon—in fact, they are well paid compared with most modern Chilean laborers. Even the lowest paid among them earns about $1,200 a month (almost triple Chile's minimum wage), and with certain under-the-table bonuses thrown in, they earn even more. Rather than squandering their wages, they use them to build the semblance of a middle-class lifestyle, complete with consumer debt, business and property loans, alimony for their ex-wives, and tuition payments for their adult children in college. A few of the men of the A shift are evangelical teetotalers, and the mercurial Mario Sepúlveda is a Jehovah's Witness who doesn't drink either. But most allow themselves a sip or two after a day's work: Whiskey, beer, and red wine are their preferred libations. A few certainly imbibe more than they should: In Copiapó, the final stop on the bus ride of the cross-country commuting southerners, one of their northern Chilean colleagues is currently drinking himself into a stupor that might keep him from going to work the next day. Working underground in modern Chile is still hard, physical labor that can leave men feeling like abused "beasts of burden," and death haunts subterranean mining today, as it always has. When Darwin rode northward he came upon the funeral of a miner, a man being carried to his grave by four of his colleagues, the pallbearers dressed in strange "costumes" consisting of long, dark-colored woolen shirts, leather aprons, and bright-colored sashes. Mining men no longer wear such dress, but in recent years the men of the San José Mine have mourned the loss of colleagues who work there. They've also seen friends maimed by the sudden explosions of seemingly solid rock that are one of the most unpredictable causes of accidents in deep underground mining. Raúl Bustos, the mechanic from the port city of Talcahuano, is a relative newcomer to the San José. But he's seen the shrine the men have built belowground to the mine's victims: Now, on the bus, he's carrying a rosary he will take with him into the mine when the workday begins.

On the last leg of their bus journey, the men enter the southern fringe of the Atacama Desert, a plain where Darwin struggled to find forage for his animals. In the Atacama, which may be the globe's oldest desert in addition to being its driest, there are weather stations that have never received a drop of rain. From their bus windows, it looks as if God had decided to pull out all the trees, and then most of the bushes and shrubs, leaving only a few hardy plants to dot the pallid brown plain with specks of olive drab. The roadside slowly comes to life again when the buses enter the irrigated, mottled greenness of the valley of the Copiapó River. Pepper trees, ubiquitous in the desert cities of the United States, are native to this corner of Chile, and they begin to appear along the roadside, dripping thin leaves onto the ground as the buses arrive in the city of Copiapó. The last five hundred yards of their bus ride takes the men past Copiapó's old public cemetery, where many generations of miners rest, including the father of one of the men of the A shift, a retired miner who drank himself to death and was buried a few days ago. After the cemetery the bus grinds quickly past a neighborhood of wood and tin hovels that is one of the poorest in the city, and then over the short bridge that crosses the channel of the Copiapó River.

The largest group of men who work at the San José live in Copiapó, the city closest to the mine. Many are veteran miners in their late forties, fifties, and early sixties, and they have pleasant memories of this riverbed. The Copiapó River was alive when they were boys, and they ran through its cooling, ankle-deep waters. Clover grew then in pools at the spot where Route 5 crosses the river, as it did when Darwin reached Copiapó and noted in his journal the pleasant aroma. A generation ago the Copiapó River began to die, and today it's become a khaki-colored eyesore garnished with trash and prickly shrubs. The average annual rainfall in Copiapó is less than half an inch and water has not flowed inside the channel since the last big storm hit the city, thirteen years ago.

When the bus pulls into the terminal, the commuting men of the A shift step out into bus bays and unload their bags. They take a short ride across Copiapó in the city's communal taxis to one of the two rooming houses where they will sleep for the next seven nights. In the last few hours before the workday begins on August 5, all the men of the A shift but one are in Copiapó or in its nearby working-class suburbs.

Geology was in its infancy when Darwin visited Chile in 1835: On his sailing journey to South America he had read one of the foundational texts of the new science, Charles Lyell's *Principles of Geology*. Once he reached Chile, Darwin witnessed a volcanic eruption in the Andes and noted the presence of seashells on ground a few hundred feet above sea level. He also lived through a two-minute-long earthquake while resting in a forest near the port of Valdivia. These experiences and observations led Darwin to deduce, more than a century before the theory of plate tectonics was formalized, that the ground upon which he was standing was gradually being pushed upward by the same forces that caused volcanic explosions. "We may confidently come to the conclusion that the forces which slowly and by little starts uplift continents, and those which at successive periods pour forth volcanic matter from open orifices, are identical," he wrote. Today geologists say that Chile sits on the "Ring of Fire," that vast seam in the Earth where continent-size chunks of the planet's crust meet. The Nazca Plate pushes underneath the South American Plate. Like a child squeezing into his bed and raising the covers into a lump, the Nazca lifted up South America to create the 20,000-foot peaks of the Andes, a process geologists call orogenesis.

The stone inside the mountains north of Copiapó was born of the magma deep inside the Earth, and is intersected with vast networks of speckled, mineral-bearing deposits. These veins were first created more than 140 million years ago, during the age of reptiles, about twenty million years after flowering plants first appeared on the planet but before the arrival of bees, and forty million years before the largest of the dinosaurs, *Argentinosaurus*, roamed the continent. A mineral-rich broth rose up through the Earth's crust, squeezing through the fissures of the Atacama Fault System for more than 100 million years, from the end of the Jurassic period to the beginning of the Paleogene. Eventually, the broth took solid form as the 200-meter-tall (656-foot-tall) cylinders of ore-bearing rock known as "breccia pipes," and also as the thin layers of interlocking veins that geologists call "stockwork." These buried deposits of quartz, chalcopyrite, and other minerals run through the hills from the southwest to the northeast, leaving lines on a prospector's map that are like an echo of the gigantic continental plates many miles farther below.

In Copiapó, two company minibuses known as *liebres* (hares) begin to pick up the men of the A shift from the two rooming houses, and from assorted stops in the city's working-class neighborhoods. There are many such buses shuttling back and forth across Copiapó this morning because these are boom times in the city, the latest in a cycle of booms and busts dating back three hundred years. A gold rush and bust in the Copiapó of the 1700s was followed by a silver rush three years before Darwin arrived. By the end of the nineteenth century the silver had run out, but the invention of nitrate-based explosives led to a saltpeter mining boom farther north in the Atacama Desert. Chilean miners provided the essential ingredient by which Europeans waged war and killed themselves in large numbers: This bonanza, in turn, led Chile to invade the neighboring nitrate-rich lands of Bolivia and Peru, with Copiapó as a base of military operations. But victory in the War of the Pacific caused yet another decline in Copiapó's economy when investment money flowed away from Copiapó into Chile's newly conquered territories. The growing global demand for copper in the twentieth century led to a new boom, however, and the building of a local copper smelter in 1951. A series of Asian economic "miracles" in the late twentieth century brought more demand for Chilean minerals and more miners to Copiapó, especially after the opening of the Candelaria open-pit copper mine in 1994. This latest boom helped to drain and finally kill the Copiapó River, because both the growing city and modern mining methods required the use of voluminous amounts of water.

In the first decade of the twenty-first century, a fourfold increase in the price of gold (to $1,200 an ounce) and record copper prices sent men deeper into the otherwise-not-so-profitable San José Mine, and into other mines in the Copiapó River valley. Copiapó's population grew to 150,000, and new and taller buildings were erected, including the tallest in the city, a fifteen-story luxury apartment building on Atacama Street, along with Copiapó's first resort, the Antay Casino and Hotel, a building whose modernist touches include a crimson, fez-shaped dome. Rising mineral prices also put more money in the pockets of the men who toiled at the San José Mine, and in recent years and months the men of the A shift have celebrated their windfall by adding rooms to their homes and by organizing large parties, often for their children and grandchildren. Sometimes these family gatherings are organized at El Pretil public park, with its watered lawns and pepper and eucalyptus trees and a small zoo with llamas, owls, and two grungy lions in a cage painted lavender.

At the end of their previous seven-day workweek, about two dozen men of the A shift attended a postwork party at the home of Víctor Segovia, a hard-drinking jumbo operator with a musical bent. A large pot was filled with beef, chicken, pork, and fish, and the host cooked up the stew known locally as *cocimiento*, a dish that itself is a kind of celebration of abundance. A few days later, Víctor's cousin Darío Segovia was planning a big birthday celebration for his baby girl on August 5, when word came that he would be needed for an overtime shift that same day (a day he was scheduled to be off). The money for this single day of work (90,000 Chilean pesos, or about $180) was too good to pass up, and he told the mother of his daughter, his companion, Jessica Chilla, that they would have to postpone the birthday party. Out of spite, Jessica refused to talk to him or to make him dinner the night before he went to work.

In the predawn hours before the shift begins, the couple have made up. At about 6:30 a.m. Darío kisses Jessica, walks down the steps that lead from his second-story living room to his front door, then stops and walks back up to wrap his arms around the woman he loves. He embraces Jessica for several seconds, a moment of gentleness and need from a brawny, callused forty-eight-year-old man. Holding her might just be his way of saying sorry, but it's also a break in their domestic routine, and it leaves Jessica anxious after Darío walks out the door.

Luis Urzúa leaves from a middle-class neighborhood in Copiapó. He's the supervisor of the A shift, and other men in his position are known to drive in their own vehicles to the mine, but Urzúa rides in with his underlings. He gets on the bus not far from another Copiapó bus stop, where he met his wife, Carmen Berríos, more than twenty-five years earlier. Urzúa is from a mining family and began working underground as a teenager, but when he met Carmen he had a prized surface job and would eventually earn a technical degree as a topographer. Carmen is a smart woman with a romantic bent, who writes poetry when the mood strikes her, and over the years she's made the hardworking Luis Urzúa her project: Among other things, she tries to get him to speak more clearly, because he often mumbles with the diction of an impoverished miner. When he finishes work at 8:00 p.m., she'll have dinner ready for him, and they'll sit alongside their two grown children, both of whom are in college.

Outside, a thick morning fog has descended over the darkened city. In a place where it almost never rains, water hovers in the air and floats underneath streetlamps, and climbs up the ravines that cut through the city. The fog is an almost daily occurrence in this part of Chile and thus has a name—*la camanchaca*. Sometimes the fog is so thick vehicles can't safely drive the highways that lead to the mine, and the start of work is delayed until it lifts, though today will not be one of those days. On street corners across Copiapó, men wait for the sound of the "hare" buses to emerge from the fog.

Each member of the A shift is, in one way or another, going to the San José this morning for the woman or women in his life: a wife, a girlfriend, a mother, a daughter. Jimmy Sánchez, who is eighteen and thus not legally allowed to work in the mine (you have to be twenty-one) has a girlfriend who is pregnant, a complication that led his relatives to beg the mine managers to give him a job. In the neighborhood named for Arturo Prat, a hero of the War of the Pacific, the smallish and handsome Alex Vega has just said goodbye to his wife, whose name is also Jessica. She's refused to give him his usual workday kiss goodbye because she's angry with him, though she will soon forget the reason. Half a mile away, in a neighborhood named for Pope John Paul II, a member of one of the crews that fortifies the mine's inner passageways leaves the home of his girlfriend. Yonni Barrios is a paunchy, soft-spoken Romeo with scarred cheeks who lives with his latest girlfriend, unless he's fighting with her, in which case he lives with his wife. Conveniently, the two women live less than a block apart, and as he walks to catch the bus he can see the door of his wife's home. He took out a loan to pay for the small neighborhood store she runs from her front door, and repaying that loan (along with helping out his girlfriend with her home) is one of the reasons he is up early today, listening for the sound of the coming bus that will take him to work.

There are many superstitions about women and mines that are expressions of the male-dominated culture's ambivalence about both women and underground labor. One legend has it that the mountain itself is a woman, and in a sense "you're violating her every time you step inside her," which explains why the mountain often tries to kill the men who've carved passageways from her stone body. Another has it that a woman working belowground is bad luck (although at least one miner has a sister who's worked for decades in her own small mine), and women are almost never seen inside the caverns of the San José. The separation of the female-centered domestic world of the home and the city from the male-centered mine in the desert is so great, most of the wives and girlfriends of the A shift workers have never been to the San José and are unaware of its exact location.

When the buses arrive, the half-asleep men take seats inside. They putter through the city and the fog, past the mustard-colored buildings of the University of Atacama near the northern edge of the city, where one of the men of the A shift has a daughter studying civil engineering. They reach the northbound stretch of the Pan-American Highway, which leads away from Copiapó toward the bones of the disappeared and the old saltpeter mines of the inner Atacama Desert. The San José is thirty-five miles outside Copiapó, and the last landmark on the drive to the mine is a rocky mountain just beyond the edge of the city known as the "Roaring Hill." Darwin saw this hill, Cerro Bramador, and wrote about the distinct noise it made. Today the sound is most often compared with that of a Latin American instrument known as a rainstick. One local legend has it that the noise is the roar of a lion guarding a golden treasure inside the mountain; another, that it's the flow of an undiscovered underground river. The quasi-scientific explanation is that the mountain's deposits of magnetite are attracting and repelling grains of sand that vibrate in the wind.

Darwin followed this road past the hissing mountain to a nearby port where the HMS *Beagle* was waiting for him; he then sailed on to the Galápagos Islands, where his observations of the local birdlife would lead him to the theory of natural selection. But the men of the A shift turn right just past the roaring mountain, off the Pan-American Highway, heading northward and inland on a narrow road of battered asphalt. For the first few kilometers, the road runs in a long, straight line across an ugly plain of grayish brown sand covered with broken rocks and windblown plants. The buses pass a cutoff to Cerro Imán, or Magnet Hill, which contains an iron ore mine, and then the road begins to curve as it enters a narrow valley surrounded by barren, rocky hills. The hills rise like reddish islands in a sea of taupe-colored sand, and shrubs the shape and size of sea urchins appear along the roadside. A man entering this landscape today still sees the relentless Atacama emptiness that Darwin saw: There are no wandering or scampering animals, nor are there any gas stations or roadside stores or any other sign of human habitation. The mountains turn maroon and orange, and resemble photographs of the surface of Mars. Finally, the buses enter another valley and come upon a blue roadside sign announcing the cutoff for the San Esteban Mining Company and its two brother mines, the San Antonio and the San José. From here, the men inside the buses can begin to see the mine's corroded and windblown structures of wood, tin, and steel, looking tragic and lonely in the alien landscape. They slowly begin a gentle climb and soon the familiar buildings on the hillside come into sharper focus: the administration bungalows, the locker and shower rooms, the cafeterias. But the men know the mine is like an iceberg city, because these surface structures represent only a small fraction of its underground sprawl.

Below the ground, the mine expands into roads that lead to vast interior spaces carved out by explosives and machinery, pathways to manmade galleries and canyons. The underground city of the San José Mine has a kind of weather, with temperatures that rise and fall, and breezes that shift at different times of day. Its underground byways have traffic signs and traffic rules to keep order, and several generations of surveyors have planned and charted their downward spread. The central road linking all these passageways to the surface is called La Rampa, the Ramp. The San José Mine spirals down nearly as deep as the tallest building on Earth is tall, and the drive along the Ramp from the surface to the deepest part of the mine is about five miles.

The San José Mine, founded in 1889, rests on top of mineral deposits that take the form of two parallel strips of softer rock embedded at a 60-degree angle inside a much harder, gray granitelike stone called diorite. An old wooden building on a mountainside marks the spot where the ore reached closest to the surface. The building once housed a working winch that lifted men and minerals out of the mine, but it hasn't been used in decades, and today it looks like a relic from an old Western. One hundred and twenty-one years after the San José Mine opened, and two thousand feet below that old building, the night shift is finishing its work during the early morning hours. Men covered in gray soot and drenched in sweat begin to gather in one of the caverns below, at a spot that is like an underground bus stop, waiting for the truck that will take them on the forty-minute drive to the surface. During their twelve-hour shift these men have noted a kind of wailing rumble in the distance. Many tons of rock are falling in forgotten caverns deep inside the mountain. The sounds and vibrations caused by these avalanches are transmitted through the stone structure of the mountain in the same way the blast waves of lightning strikes travel through the air and ground. The mine is "weeping" a lot, the men say to each other. *"La mina está llorando mucho."* This thundering wail is not unusual, but its frequency is. To the men in the mine, it's as if they are listening to a distant storm gathering in intensity. Thankfully their shift is about to end. A few will tell the next group of men to enter the mine, the men of the day A shift, *"La mina está llorando mucho,"* but it is unlikely the San José will close as a result. The men who work there have heard these gathering storms before. The thunder always recedes and eventually the mountain returns to its steady and quiet state.

As the men of the A shift reach the mine property, they pass a guard shack and then the entrance to the Ramp, a tunnel first blasted from the hard diorite rock of the mountain more than a decade ago. The opening to the San José Mine is an orifice five meters wide and five meters tall, and the edges that face the outside world resemble a series of stone teeth. Trucks filled with men and ore now begin to emerge from this mouth, because the prior shift is finishing its workday. They've removed a few hundred tons of ore-bearing rock, with copper sulfide in fingernail-size specks that glimmer with the same marbled pastels of art nouveau paintings: crimson, forest green, maroon, and the brassy yellow, tetragonal crystals of the copper ore known as chalcopyrite. When processed, each metric ton of this rock produces as much as forty pounds of copper (worth about $150), and less than an ounce of gold (worth several hundred dollars). The gold is invisible, though many older men of the A shift grew up listening to their fathers say you could taste it in rock like this.

The men file into changing rooms that are as moldy and cramped as those on an old sea vessel. They put on overalls, attach freshly charged battery packs to their belts, and lamps to their blue, yellow, and red helmets. Luis Urzúa puts on the white helmet that's a symbol of his managerial status, and also straps a palm-size "self-rescue" oxygen canister to the leather belt on his waist. Urzúa is a relative newcomer to the San José, easygoing for a shift supervisor, and doesn't know his crew as well as he would like, in part because the crew is always changing from one day to the next. Today, one man will join the A shift and work underground for the first time; and as Urzúa enters the mine he notes that another of the two dozen or so men working for him today hasn't even made it to work yet.

After his long journey from Santiago, Mario Sepúlveda has arrived in Copiapó too late to catch the minibuses to the mine. He stands on a Copiapó street corner and thinks that this is a good thing. The last time he was in town he talked to a friend who runs another small mine. This friend was well aware that the San José Mine is in a perpetually precarious financial and structural state, and offered Sepúlveda another job. Now it's 9:00 a.m., the day A shift at the San José started an hour ago, and Sepúlveda thinks that if he's lucky they'll fire him at the San José for not showing up today, which will make it easier for his friend to hire him at that other mine. These thoughts are running through his head when the driver of another minibus passes by and spots him.

"Perri!" the driver calls through the window, using Sepúlveda's nickname. "You missed your bus? I'm going that way. I'll give you a ride. Jump in."

The man with the heart of a dog arrives at the San José Mine after 9:30, more than ninety minutes late. The fog has burned off by then, and Sepúlveda stands in the desert sunlight for a final few moments before getting a ride down into the mine and his workplace.

PART I

BENEATH THE MOUNTAIN OF THUNDER AND SORROW

1

A COMPANY MAN

In the San José Mine, sea level is the chief point of reference. The five-by-five-meter tunnel of the Ramp begins at Level 720, which is 720 meters above sea level. The Ramp descends into the mountain as a series of switchbacks, and then farther down becomes a spiral. Dump trucks, front loaders, pickup trucks, and assorted other machines and the men who operate them drive down past Level 200, into the part of the mountain where there are still minerals to be brought to the surface, working in passageways that lead from the Ramp to the veins of ore-bearing rock. On the morning of August 5, the men of the A shift are working as far down as Level 40, some 2,230 vertical feet below the surface, loading freshly blasted ore into a dump truck. Another group of men are at Level 60, working to fortify a passageway near a spot where a man lost a limb in an accident one month earlier. A few have gathered for a moment of rest, or idleness, in or near El Refugio, the Refuge, an enclosed space about the size of a school classroom, carved out of the rock at Level 90. As its name suggests, the Refuge is supposed to be a shelter in the event of an emergency, but it also serves as a kind of break room because fresh air is pumped into it from the surface, offering a respite from the humidity and heat, which often reaches 98 percent and 40° Celsius (104° Fahrenheit) in this part of the mine. The San José is said by the men who work there to be like hell, and this is a description with some basis in scientific fact, since it's the geothermal heat emanating from the bowels of the Earth that makes the mine hotter the deeper they go.

The mechanics led by Juan Carlos Aguilar find respite from the heat by setting up a workshop at Level 150, in a passageway not far from the vast interior chasm called El Rajo, which translates loosely as "the Pit." Air circulates through the Pit and the faintest hint of a breeze flows from that dark abyss into the makeshift workshop. The mechanics have decided to start their workweek by asking Mario Sepúlveda to give them a demonstration of how he operates his front loader. They watch as he uses the clutch to bring the vehicle to a stop, shifting from forward directly to reverse without going into neutral first.

"Who taught you to do that?" the mechanics ask. "That's wrong. You're not supposed to do it that way." He's mucking up the transmission by doing this, wearing out the differential.

"No one ever showed me," Sepúlveda answers. "I just learned from watching." The mechanics work for a company that contracts maintenance services to the mine, and they are not surprised to learn that an employee of the San José is operating an expensive piece of equipment without having received any formal training. The San José is an older, smaller mine known for cutting corners, and for its primitive working conditions and perfunctory safety practices. Among other things, it has vertical escape tunnels that will be useless in an emergency because they lack the ladders necessary for the miners to use them.

Newly informed as to the proper use of the clutch, Sepúlveda leaves the mechanics to work down at Level 90.

Throughout the morning, the mountain has continued its intermittent thundering wail, the sound of a distant explosion followed by a long whining sound. Carlos Pinilla, the general manager of the San Esteban Mining Company, hears this noise as he travels in a pickup truck between the levels of the San José Mine. He has an office on the surface, but is now deep inside the mountain to impose some discipline on a workplace that's much too casual for his liking. "I had to reprimand everyone from the shift supervisor on down," he says. "None of these guys was a little white dove. I didn't want them to be afraid of me. But if I'd go down there and find six guys sitting around chatting, I wanted them to at least stand up when they saw the boss. Without that, everything would start to fall apart . . ."

Pinilla is a jowly man of about fifty who's worked his way up from lowly office jobs in mining companies to one in which he's the general manager of the two mines run by the San Esteban Mining Company. He's described by his underlings as imperious, the kind of a man who will bark an order and who treats the miners as if their sweating, helmeted presence were offensive to him somehow. In a country of rigid class distinctions, such as Chile, laborers are often subjected to bald conde-scension by the salaried classes. Even in this context, to the miners Pinilla stands out as an especially domineering "white helmet," particu-larly in contrast to the soft-spoken white helmet beneath him in the mining hierarchy, the shift supervisor Urzúa. In recent weeks, one of the members of the A shift, Daniel Herrera, had asked Pinilla several times for replacement air filters for the masks the men wear, until, he claims, the general manager finally replied, sarcastically: "Yeah, I'm going to get you a whole truck filled with filters!" Pinilla is *"el amo de la mina,"* the miner Jorge Galleguillos says, lord and master of the mine. Galle-guillos is fifty-six, and older men like him are afraid of Pinilla because he can fire a man in an instant, leaving him in the unenviable position of looking for work in an industry where youth and a stout constitution are especially prized. At the same time, it's only the older, most experi-enced miners who have dared to speak out in the face of the mounting evidence of the San José Mine's structural weakness.

After 121 years in which men and machines have emptied and hol-lowed the mountain, the San José Mine is still intact thanks to the hard, gray diorite stone that makes up most of the mountain's mass. In mining slang, the diorite is "good" rock in the sense that it holds together when you drill through it. If the ore-bearing rock is like a crumb cake that begins to disintegrate as soon as you poke it, the diorite is more like a stiff custard. Generally speaking, the diorite provides an excellent, sta-ble structure for a tunnel, requiring relatively little reinforcement. The Ramp has been carved through this stone, and is the only true way in and out of the mine. Until recently, no one who works in the San José believed it was in danger of collapsing. Then, several months back, a finger-wide crack was discovered in the Ramp at Level 540.

Mario Gómez showed the crack to his shift supervisor as soon as he saw it. Gómez is a sixty-three-year-old miner who drives a thirty-ton-capacity truck into the mine. "I'm pulling my truck out of this mine,"

Gómez said then. "And I'm not going back in, and no one else will, until you get the mine manager and the engineers here from Copiapó, and make them look at this and evaluate it." A few hours later, the engineer and the general manager arrived. They placed mirrors inside the half-inch-wide crack: If the mountain was still shifting and splitting, then the movement would break the mirrors. But the mirrors are still intact.

"Look, the Ramp is the safest thing in this mine," the manager said to the miners. "All that cracking is coming from the Pit. The walls of the Pit can collapse up to five meters away and nothing will happen to the Ramp." More mirrors were placed in the crack when water began to leak through it, but they all remained in place and intact for weeks and months. Galleguillos studied the mirrors every time he drove past one. He wrote down other troubled observations in a notebook: "Falling material is felt at Level 540 . . . tunnel walls detached at Level 540." Then he forced the mine manager to sign a copy of these notes. Later, he confronted the manager again.

"How do we know you're not going in and replacing those mirrors when we're not looking?" Galleguillos asked.

"What are you?" the manager snapped back. "A coward?"

Now Pinilla crosses paths several times with the workers as he patrols the interior of the mine in his pickup truck. At midmorning, Yonni Barrios and his crew of fortifiers see him at Level 60, and tell him the mountain is making noises you ordinarily can't hear that far down. "Don't worry," he tells them, "the mountain is just settling" ("*el cerro se está acomodando*"). Higher up, at Level 105, another group of workers has a similar conversation. The thunder can be heard in every corner of the mine, and it's causing a sense of worry to spread through the passageways—and also a sense of denial. Mining is an inherently dangerous occupation, and those who have decades of experience working underground take pride in facing its risks. The men of the A shift have made it a habit of complaining to their wives and girlfriends about the San José, using the preferred euphemism that conditions in the mine are "complicated," and then brushing off the danger when pressed for details.

Luis Urzúa, too, has told his wife the San José is "complicated," and when he took the job there a few months ago, it was in full knowledge of the mine's recent accidents. This morning he's hearing the complaints about the thunder from his crew, including a few who insist they should all go to the surface. Urzúa says to wait. Urzúa is fifty-four years old, and despite his degree as a mining topographer, he freely admits that he's intimidated by people who are "bigger" than he is. He could confront his boss, Pinilla, and demand that all his men be pulled out: In fact, a few of the men of the A shift are starting to think he's weak for failing to do so. But at this moment none of those men complain very loudly either, nor do they announce that they will simply refuse to work a minute longer and leave the mine immediately, a step men at the San José have taken before.

Mario Gómez, the oldest man in the A shift, has two missing fingers on his left hand as a reminder of what can happen underground from one moment to the next. At about 12:00 p.m., he, too, is given a warning of the impending disaster: There's "smoke" coming from Level 190, the driver Raúl Villegas tells Gómez as the two men and their dump trucks pass each other on the Ramp. But Gómez listens to the tough, gruff voice in his head that tells him he should be careful but not fearful. When he drives past the "smoke" and takes a look he concludes: *It's just dust, and dust is normal in the mine.*

Still, the reports of unusual noises and explosions keep coming, and by late morning the boss of all the bosses, Carlos Pinilla, is, according to several of his underlings, starting to act strangely. Urzúa and his second-in-command, the foreman Florencio Avalos, spot him in his pickup at Level 400. Pinilla stops to shine a large flashlight at the stone walls of the Ramp, and another worker who spots him at this moment says: "His flashlight was huge, so much bigger than the ones we carried, that it made me nervous to see him using such a thing." Later, other workers see him going into one of the corridors that lead off the Ramp to shine the same flashlight inside the cavernous, excavated space of the Pit. They also see him standing next to the pickup truck, as if he were listening for something, or trying to feel movement inside the mountain. He seems to be listening, too, when he stops at the entrance to corridors near Level 400, to wipe clean the blue-and-white placards reading "DO NOT ENTER" and "BYPASS." "I thought that was strange," Urzúa says, "to see my *jefe* cleaning the traffic signs." When Florencio Avalos comes upon

Pinilla a little bit after noon, the general manager tells him he has a flat tire and needs to get a spare as quickly as possible. "He seemed nervous," Avalos says. "As soon as we changed the last bolt, he took off, and we never saw him again."

As Pinilla drives toward the surface at around 1:00 p.m., he crosses paths on the Ramp with Franklin Lobos, a tall, balding onetime soccer star and minor local celebrity whose chief fame underground is that he's a grouch. Lobos drives the personnel truck that ferries the men in and out of the mine, and at that moment he's headed down to pick up the men for lunch.

"Franklin, I'd like to make two observations to you," Pinilla says. "First, I want to congratulate you, because you have the Refuge nice and clean." The Refuge is stocked with two metal cabinets with food, supposedly enough to keep an entire shift alive for two days. As the driver of the personnel truck, Lobos holds the keys to those cabinets and is responsible for keeping the Refuge in order. "And second," Pinilla continues, "as soon as you can, I want you to go and talk to the supply master. Because we're getting a box ready with more provisions for the Refuge." More food, blankets, and a first-aid kit, that sort of thing, Pinilla says.

Pinilla seems to be in a hurry to leave the mine, and is making preparations for an emergency, but he says it isn't because he thinks the mine is about to collapse. His biggest concern isn't an accident in the mine, but rather that the Chilean government agency in charge of mining safety will close down the mine. The big flashlight was required for a routine inspection of the excavated cavern of the Pit: Given that it's as much as six hundred feet tall in places, he needs an especially powerful beam. And he's just ordered that more emergency supplies be taken to the Refuge because the miners are always stealing the food inside (he'd finally bought a lock and an aluminum band to keep them from doing so), and if an inspector finds the supplies lacking he'll shut down the mine.

In 2007, the Chilean government had ordered the San José Mine closed after an underground explosion of rock killed the geologist Manuel Villagrán. The mine owners promised the government they would take a series of steps to improve safety, and the San José was allowed to be reopened. (The men built a shrine of votive candles to Villagrán at the spot where he died and where the vehicle he was driving remains buried.) Unlike other mines in the Copiapó region, the San José is not owned by a big foreign conglomerate, but rather by two sons of the late Jorge Kemeny, an exile from Communist Hungary who settled in this part of Chile in 1957. Marcelo and Emérico Kemeny did not, unfortunately, inherit their father's passion and skill for the mining business. Emérico left his share of the work of owning the mine to his brother-in-law, Alejandro Bohn. The Kemenys and Bohn struggled to keep the San Esteban Mining Company profitable while complying with the Chilean government's demands. The company was required to install ladders inside the ventilation tunnels as an alternate emergency escape to the Ramp, and also new fans to increase air circulation and lower the temperature at the bottom of the mine, which at times neared 50° Celsius (122° Fahrenheit). As if in recognition of their previous shortcomings on safety, the owners then contracted with a company called E-Mining to take charge of the mine's daily operations. E-Mining recommended a seismic monitoring system designed to detect potentially catastrophic shifting of the mountain's internal structure; it was never purchased. The contractor also recommended other movement-detection devices known as geophones, but after a month they stopped working because the mine's trucks kept running over their fiber-optic cables. Eventually the San Esteban fell behind in its payments to E-Mining, and the company canceled its contract and withdrew its employees. The San Esteban then hired a former San José employee, Carlos Pinilla, to take the contractor's place. The San Esteban company doesn't have enough money to pay for the seismographs, or to keep the geophones operational, nor has it installed the ladders or the ventilation systems the government ordered. Basically, it's impossible to do those things and still keep their medium-size mine profitable. Among other things, the company is $2 million in debt to ENAMI, a government-owned company that processes ore for small- and medium-size mines. Just like the workers who know how dangerous the San José is and work there anyway, the owners know how dangerous it is and keep it open nonetheless. To keep the company and its financial ambitions and responsibilities afloat, they're gambling with the miners' lives.

As he drives to the surface, Carlos Pinilla is doing exactly what the owners have always asked him to do: keep the mine running, with the cash-producing ore coming out, while cutting corners to keep costs down, hoping for the best, trusting that the very hard diorite of the Ramp will hold together and allow the men to escape, even if the internal structure of the mountain, weakened after more than one hundred years of digging and blasting, causes the rest of the mine to crumble.

If Pinilla closes the mine and orders everyone out, and the mine doesn't collapse, it might cost him his job. And besides, he believes at this moment that the San José Mine has at least another twenty years left in it.

Just after 1:00 p.m., two men cross paths on a road carved out of stone: One is headed up, the other down. Carlos Pinilla, the man in the white helmet who worked his way up from warehouse clerk to general manager, revs the engine of his pickup to begin to climb to the surface and daylight. Franklin Lobos, a man whose fortunes have been in a precipitous decline, watches from beneath a blue helmet as the *jefe* drives away. Lobos reaches over and releases the emergency brake on his truck, so that gravity does the first bit of the work of sending him on his journey downward. He turns on the fog lamps—the main beams of the truck's headlights have never worked—and heads down to the Refuge, below Level 100, where the men are starting to gather and wait for him for the ride up for lunch.

Descending a bit more, at Level 500, Lobos sees a truck coming up, and since uphill traffic has the right of way, he lets it pass: It's Raúl Villegas, the driver who's just complained about "smoke," driving a big dump truck filled with tons of ore.

The men wave hello and goodbye to each other and soon Lobos is on his way down again. He reaches Level 400, where the signs gleam a little brighter thanks to Pinilla's polish job. The older miner Jorge Galleguillos is riding in the cab alongside Lobos, going down to check on the system of tanks and hoses that brings water from the surface down into the mine. The drive is slow and tedious, following the ground-hugging beam of the truck's fog lamps along a single gray tunnel, sinuous and repetitive, as if they were entering the dark, dank, and vacant landscape of a miner's subconscious. A half hour longer they drive, one rocky turn following another, in passageways with a million ragged, serrated edges blasted from rock. They are at about Level 190 when they see a white streak move past the truck's windshield from right to left.

"Did you see that?" Galleguillos says. "That was a butterfly."

"What? A butterfly? No, it wasn't," Lobos answers. "It was a white rock." The mine's ore-rich veins are thick with a translucent, milky quartz that glimmers when it catches the light.

"It was a butterfly," Galleguillos insists.

Lobos believes that it's pretty much impossible to think that a butterfly could flutter down in the dark to more than a thousand feet below the surface. But for the moment he surrenders the argument.

"You know what, you win. It was a butterfly."

Lobos and Galleguillos continue driving for about twenty more meters. And then they hear a massive explosion, and the passageway around them begins to fill with dust. The Ramp is collapsing directly behind them, near the spot where a rock or a butterfly passed before their windshield.

The sound and the blast wave interrupt thirty-four men laboring inside stone corridors. Men using hydraulic machines to lift stone, men listening to stone crash against the metal beds of dump trucks, men waiting for the lunch truck in a room carved from stone, men drilling into stone, men driving diesel-fed machines down a stone highway, and men wearing eroded stone on their clothes and their faces.

The truck driver Raúl Villegas is the only one of the thirty-four men underground at the moment of the collapse who manages to escape. He watches in horror as a dust cloud gathers in his rearview mirror and quickly overtakes his truck. He speeds through the cloud toward the exit, and when he reaches the mouth at which the Ramp opens to the surface, the dust follows him outside. A gritty brown cloud will continue flowing out of that malformed orifice for hours to come.

Inside the personnel truck at Level 190, Lobos and Galleguillos are the two men closest to the collapse, which hits them as a roar of sound, as if a massive skyscraper were crashing down behind them, Lobos says. The metaphor is more than apt. The vast and haphazard architecture of the mine, improvised over the course of a century of entrepreneurial ambition, is finally giving way. A single block of diorite, as tall as a forty-five-story building, has broken off from the rest of the mountain and is falling through the layers of the mine, knocking out entire sections of the Ramp and causing a chain reaction as the mountain above it collapses, too. Granitelike stone and ore are knocked loose, pulled downward to crash against other rocks, causing the surviving sections of the mine to shake as if in an earthquake. The dust created and propelled by the explosions shoots sideways, upward, and downward, ejected from one passageway and gallery in the mine's maze of corridors to the next.

In an office about one hundred feet above the mine opening, Carlos Pinilla, the hard-driving general manager, hears the thunder crack and his first thought is: *But they're not supposed to be blasting today.* He concludes that it's probably another collapse of rock inside the Pit, which is nothing to be worried about. But the sound of rolling thunder doesn't stop. His phone rings, and the voice on the line says, "Step out your door and look at the mine entrance." Pinilla walks into the midday sun and sees a billowing cloud of dust bigger than any he's seen before.

2

THE END OF EVERYTHING

When the diorite "mega-block" first breaks loose, it emits a loud, percussive sound. Many of the thirty-three men trapped below will not hear it, because they're wearing ear protection deeper in the mine, or because they're operating loud, heavy machinery. The mechanics at Level 150 are working on a twenty-seven-ton Toro 400 "load-haul dumper," parked just thirty feet or so from the precipitous drop-off where the Pit begins. They're behind schedule, and trying to finish before lunch, because they've spent ninety minutes waiting for a special wrench that someone had to go up to the surface to fetch. At 1:40 p.m., three men are using that tool to tighten the last two bolts on one of the squat machine's five-foot-tall wheels when they hear what sounds like a gunshot. A moment later, they are knocked off their feet by a blast wave, and then enveloped by the sound of falling rock, and the walls around them begin to shake, and stones the size of oranges are falling around them. Raúl Bustos, who survived an earthquake and tsunami five months earlier, scurries under the chassis of the Toro 400. So does Richard Villarroel, a twenty-six-year-old whose young girlfriend is six months pregnant with their first child. Juan Carlos Aguilar, the man who's traveled here from the rain forests of southern Chile, grabs on to a nearby water pipe. For two minutes, their ears are filled with the sound of the mountain's collapse: One will describe it as like hearing many jackhammers taking apart a sidewalk all at once. Then a second blast wave, going in the opposite direction from the first, sweeps through the corridor, causing more rocks to fall. Stones falling from inside the nearby cavern begin to fill their informal workshop, and when the noise and the crashing sounds finally ease a bit, they look and see that one of the other vehicles closer to the edge of the cavern is half-buried in rock.

The three men gather themselves and in shouts agree to head on foot to the Ramp, to search for the fourth member of their work crew, a man who is driving a pickup truck they might use to escape.

A few minutes earlier, Juan Illanes had been ordered by his boss, Juan Carlos Aguilar, to leave the informal workshop to go get some drinking water. This requires a short drive down to the Refuge in a pickup truck. Illanes is driving a bit past Level 135 when he comes upon a slab of rock that's fallen. It's about six feet long, and maybe ten inches thick, too big to drive over. He has to get out and move it, or find a way around it. So he puts the pickup in reverse, and is just taking his hand off the gearshift when he hears a powerful "rocket blast" (*un cohetazo*). The walls of the Ramp begin to spit small stones and he hits the accelerator to go back up, in reverse, but only for a second or two, because then he feels the blast wave hit his truck, followed by a dust cloud, and an earthquake, as if the Ramp were inside a cardboard box that someone had decided to shake, he later says. He waits a bit and then, remembering his coworkers back at Level 150, he turns the pickup around and drives back up the Ramp toward the workshop, steering the vehicle into the angry cloud of oncoming dust. The dust is so thick he can't see where he's going, and he keeps crashing into the side wall of the tunnel. Finally he stops, and waits, with the lights on and the motor still running. He's sitting there, inside the cab, when a figure steps out of the dust cloud and into the short beam of the truck's headlights—it's his boss, Aguilar. Two more men are running toward him, the other mechanics, Raúl Bustos and Richard Villarroel.

Illanes tells them it's impossible to drive the pickup truck through the dust, so the four men find a section of the Ramp where there's a wall that's been fortified with steel mesh. They huddle against this wire net, the flimsiest of defenses against a mine that's crumbling all around them.

•

The sound and the blast wave continue to race downward through the mine, past another group of workers at Level 105 who are perforating rock, including a forty-seven-year-old widower named José Ojeda. At Level 100, Alex Vega is waiting for Franklin Lobos and his lunch truck, taking a break, chitchatting with other workers, including Edison Peña, a thirty-four-year-old Santiago native and mechanic with a reputation at the San José Mine as a troubled and anxious individual. Peña keeps physically fit by riding back and forth across Copiapó on a bicycle he's named Vanessa, after a porn star whose athleticism he admires. As the lunch hour approaches, Peña is depressed because he had been hoping that with all the commotion about the mountain making noises, the rest of the workday would be canceled, but it hasn't been. Both men hear the thunder blast from above, a noise that's dampened a bit after traveling through a few hundred feet of granitelike stone to reach them. "We were used to noises," Alex later says. "It was normal for the mountain to crunch and crash. Like the boss would say to us: 'The mountain is a living thing.'" But this noise is unlike any other they've heard before, and it's followed by a rumbling sound that grows in volume with each passing second. The resting miners in blue, yellow, and red helmets standing with Vega and Peña at Level 105 begin to look around, and to look at one another. Their faces ask, *Does anyone know what that is?* Finally, someone shouts, "*La mina se está planchoneando.*" The mine is pancaking. There is a burst of wind, and then they see a cloud of dust flowing into the Ramp from tunnels leading to old, no-longer-worked sections of the mine. The cloud races down the Ramp toward them, and soon it is upon them, showering them with dirt and pebbles as they run downhill for the safety of the Refuge.

About ten vertical yards below Vega and Peña, Samuel Avalos is among a second group of workers waiting for the lunch truck, gathering inside or near the Refuge itself. Avalos's recent past is even more unsettled, his work history more scrappy, than that of most of his fellow miners. Until recently, he worked as a street vendor, and he's still hawking pirated discs as a side job, and has thus earned the nickname "CD." He's a small, funny man, at once shy and possessed of an ironic sense of humor that's allowed him to endure a life in which he also sold flowers and other wares on street corners to survive. Once inside the Refuge, he performs a midday ritual that would seem eccentric or lunatic in any other workplace: He strips down to his underwear while several of his fully clothed coworkers stand by. After just half a shift of work, his overalls are sweat-soaked, so he takes them off, wrings the moisture out, and hangs them on one of the water pipes in the Refuge to dry. The Refuge is a room carved out of stone, with a white tile floor, a cinder-block wall, and a steel door that separates it from the Ramp. Other men come in and go out, but no one says anything to Avalos as he stands there half-naked, relaxing. Avalos has been waiting at the Refuge for quite a while, long enough for his uniform to be a bit drier, and he's putting the overalls back on when he hears the thunderclap.

At first Avalos wonders if someone is blasting rock in the mine: but there is no blasting scheduled for that day. It's bad if someone is setting off explosions without warning the men working in the mine, though it's been known to happen. At the same time, the sound is louder than any controlled explosion. What could it be?

The curly-haired Víctor Zamora, who grew up in the town of Arica on the Peruvian border, is near the Refuge, too, sitting on a stone that's like a bench. A chain-smoker, he's puffing on another cigarette and feeling pretty good, pretty relaxed, despite the dull, sometimes throbbing pain of a molar that's got a cavity. He's there with the members of his fortifying crew, and later he will remember this moment as one of contentment and brotherhood. The men who work together in any given shift refer to one another as *los niños*, whether they be twenty-one or sixty-one, and Zamora actually likes working in the San José Mine with his fellow "boys": He feels comfort in the way men treat one another as equals, "no one better than anyone else." Now Zamora hears the explosion, too, and since he's worked in mines on and off since he was a young man, the first thing he does is nothing more than take another drag on his cigarette.

A minute or so later, the first blast wave reaches Zamora and knocks him off his stone bench and violently throws open the heavy metal door to the nearby Refuge. He rises to his feet and runs inside.

In the panicked minutes that follow, Alex Vega, Edison Peña, and several other miners who had been waiting for the lunch truck nearby run into the Refuge to join Samuel Avalos, Víctor Zamora, and the others there. Soon, about two dozen men have huddled inside. Outside, beyond the steel door, the mountain is caving in upon itself. After fifteen or twenty minutes, the noise isn't as loud, and then men drum up the collective courage to make a run for safety, heading out past the steel door and the cinder-block wall, stepping onto the Ramp to walk, scurry, and run to the surface, nearly four miles away.

After a morning spent driving to every level of the mine where men are working, the shift manager, Luis Urzúa, is at Level 90, not far from the spot where Mario "Perri" Sepúlveda is operating a front loader. At 1:40 p.m. he hears a loud crash that's audible over the roar of the front loader, what sounds like a huge rock falling in the Pit. It's normal to hear such noises coming from the mine's excavated caverns when workers have been extracting ore at higher levels, and Urzúa isn't especially concerned. Five minutes later, he hears another crash, and he tells Sepúlveda to stop his machine, but Mario is already doing so, because he's just felt what he thinks is one of the large tires on his rig blowing out. Sepúlveda is taking off his ear protectors when the pressure wave passes through the tunnel and plugs up his ears. *What is this,* he wonders, and then Florencio Avalos pulls up in the shift manager's white Toyota Hilux pickup truck to announce that the mine appears to be collapsing. Sepúlveda and Urzúa jump into the pickup and the three men drive up the Ramp toward the Refuge, where they find that all the men that should be there, seeking shelter from the collapsing mine, are gone.

With Sepúlveda still in the pickup with them, the two supervisors, Urzúa and Avalos, circle back downhill. They drive away from the surface and safety because there are men deeper in the mine. "We have to make sure those *huevones** get out," he says. Getting each and every man working underground out at the end of the day is the responsibility Urzúa assumes each time he enters the mine.

---

*A common Chilean insult and term of endearment derived from *huevo,* or egg, a slang word for testicle.

•

Thirty vertical meters farther down, at Level 60, the Bolivian immigrant Carlos Mamani is at work inside another front loader. These are his first hours underground at the mine: He passed an aboveground test to operate the loader a few days before, and that very morning he'd taken a kind of final underground exam, supervised by one of the mechanics. After watching him awhile, the mechanic left Mamani to work alone for the first time. It's the kind of moment Mamani, an earnest, baby-faced twenty-four-year-old, had imagined for many years. He grew up speaking Aymara on a rural farm on the desolate, beautiful Altiplano, one of the poorest corners of the poorest country in South America. As a teenager he joined the immigrant stream to Chile to work picking grapes and construction, all the while secretly dreaming of being a police detective. Instead of going to college he settled for a bit of technical training, and today he's soloing at the controls of a big, Swedish-built machine with a joystick and digital gauges for the first time. The gauges glow in the darkness of the cavern, adding to the sense that he's operating a complex and modern piece of equipment.

Mamani's loader is attached to a big basket, with two men suspended inside, working on the roof of a tunnel, using jackhammer drills to do what's called "fortifying." The four-man crew working with him includes Yonni Barrios, the man with two households, and Darío Segovia, the man who gave his wife an unusually long and heartfelt embrace that very morning. They're drilling six-foot-long metal rods into the stone to hold up a steel mesh that's meant to keep any slab from falling on the people working below. Mamani doesn't know it, because it's his first day, but the miners inside the basket do: This is the spot where one month earlier the miner Gino Cortés was crushed by a falling slab of rock weighing more than a ton, causing his left leg to be amputated. Mamani has forgotten to bring down the lamp he was issued—he left it in the locker room up above, but Barrios told him not to worry, he can go pick it up at lunchtime. Mamani doesn't even know what the daily routine of the mine is supposed to be, and he's looking at his watch, and seeing it's almost 2:00 p.m., he's wondering: *When are we going to eat? When are these guys going to stop working? Are they ever going to stop working? Next time, I'll have a bigger breakfast.*

Above him, the men stop working suddenly. Through the glass of the cab, Mamani watches them. Barrios and Segovia are looking at themselves as if to say, "What was that?"

Barrios is operating a jackhammer and has earplugs in, and hasn't heard the distant explosion. Instead, he's just felt a kind of pressure wave pass through his body. He feels his entire body being squeezed, and then a bit later, unsqueezed, as if he were inside the cylinder of a hand pump while someone pushed down on the plunger and then pulled it up again. Segovia has heard the explosion, and looks down at the two men standing below them, their assistants Esteban Rojas and Carlos Bugueño. "Something is happening," one of them shouts. "Our ears are all plugged up." But the crew keeps on working for several minutes more, until pebbles begin to fall from the roof of the tunnel, and dust blows inside.

Inside the cab, Mamani wonders if the gritty cloud beginning to build outside his front loader is somehow part of the work, too. But no, it isn't, because the miners begin gesturing for him to lower the basket, and to back up the loader out of the tunnel, quickly. Mamani does this, then turns and sees Daniel Herrera reaching for the door to the cab. When Herrera opens the door, Mamani's ears plug up suddenly and he loses much of his sense of hearing: He sees the lips of people moving but struggles to make out the words.

One of the workers starts to move his flashlight in circles, a signal that Mamani knows, from other mines he's worked in, means something very frightening: *Get out! Evacuate the mine! Get out now!* From where they find themselves, it's a five-mile drive to the surface, a vertical climb of two thousand feet, some forty minutes going downhill, and God knows how long uphill, because it's a drive Mamani has never taken.

Moments later Mamani is driving the loader onto the Ramp, carrying several men into a tunnel filling and then filled with dust, headed for the Refuge, the big machine hitting the wall because he can't see.

A little way forward, at Level 90, they see the shift supervisor, Urzúa, and the foreman, Avalos, coming down. Keep going up, the bosses say. We'll catch up to you. We're going down to look for the two men working deepest in the mine.

·

During the course of morning, Mario Gómez has made three trips down into the mine, and then back up to the top. Going downhill with an empty truck, he can do it in about thirty minutes; back up, fully loaded with gold- and copper-laden rock, it's more than an hour, grinding the engine in first and second gears. Just after noon, he's at the top with an empty truck and decides to take his lunch break. He enters the corrugated-metal company cafeteria and puts the container of rice and beef he's brought and sticks it into the microwave, then takes it out: After just one spoonful, he stops. Gómez is paid a base salary, but also a set fee for each trip down into the mine. He thinks about the money and decides he should probably eat his lunch down at the bottom while the loader is filling up his truck, and thus squeeze in an extra trip for the day. The additional trip, which will nearly cost him his life, is worth 4,000 Chilean pesos, or about $9.

Gómez gets into his cab and drives down to Level 44, where he parks inside a corridor with piles of ore-laden rock waiting to be carried to the surface. He is, at that moment, the man working deepest in the mine, some 2,218 vertical feet from the surface. The man operating the loader that's supposed to lift the ore into his truck isn't there, so Gómez starts eating his lunch in the cab, with the engine running and the air-conditioning on against the 100-plus-degree heat. Ten minutes later the loader operator arrives: white-haired, fifty-six-year-old Omar Reygadas. He lifts one scoopful of rock and tosses it into the bed of the truck. At that moment, Gómez feels a puff of air against his face, which is odd, because the windows of the truck cab are all closed. Then he feels a burst of pressure between his ears, as if his skull were a balloon being inflated, he says. The truck's engine stops, and after a few seconds it starts again, all by itself. All the while Reygadas keeps working his loader, and the crash-grind made by stone hitting the truck's metal bed leaves both Gómez and the loader deaf to any other sound. Reygadas has felt the rumbling and the pressure wave, too, and thinks that the shift supervisor, Urzúa, has ordered some blasting without bothering to tell everyone. It's another dangerous screwup in this already screwed-up mine, and the last straw for Reygadas. *That's it,* he tells himself: *I'm going to finish this job and go cuss out that idiot Urzúa and tell him I quit.*

When Gómez's truck is fully loaded, he begins his drive upward to the surface. But he advances only a few hundred feet or so up a steep section of the Ramp when the tunnel around him begins to fill with a dust cloud. This isn't especially worrying, because he's seen it happen before, and he tries to push the truck through the cloud, but it gets so thick he can see only a few feet in front of the windshield. Gómez is in danger of crashing against the wall, so he stops and opens the door and feels the wall—it's straight, not curved, and he gets back in the cab and points the steering wheel straight and goes faster, driving blindly until Urzúa appears next to his window, gesturing for him to stop and get out. Gómez lowers the window, and at that moment he is assaulted by a deafening noise, the memory of which will haunt him in the days, weeks, and months to come, causing him to weep when he remembers it: He hears the rumble of many simultaneous explosions, the sound of rock splitting, the stone walls around him seeming to crack, as if they might burst open at any moment.

The men who were at the Refuge try twice to escape on foot during lulls in the explosions. After a first attempt ends with a retreat back to the Refuge, they try again, only to find the rumbling of an underground earthquake beginning anew. The solid rock of the mountain is transformed into a breathing, pulsating mass. The ceiling and floor of the Ramp become undulating waves of stone, and the mountain hurls boulders that emerge from the blackness of the tunnel and roll and bounce downhill, each a lethal weapon aimed at their bodies. "We were a pack of sheep, and the mountain was about to eat us," José Ojeda later says. For Víctor Zamora, the sound of exploding rock feels like machine-gun fire aimed at him and his fellow miners. It's too much, too scary, too dangerous, so they start running back downhill, but it's as if they were running on a bridge swaying in the wind, one of the miners says. Luis Urzúa and Florencio Avalos arrive at this moment, and see this group of panicked men running toward their pickup truck. They watch, mesmerized, as another blast wave rushes through the tunnel. It seems to pick up Alex Vega, the smallest and slightest of the miners, and lifts him off his feet, as if he were some miner-shaped kite that caught a sudden gust of wind. Others are knocked over, falling, flailing at the air. They stumble, these big men in overalls, men with bodies shaped by red wine and beer and backyard barbecues, babied by their wives and mothers and mothers-in-law. The blast knocks Zamora against the wall of the Ramp, face-first, knocking out some of the teeth he was born with and a few others that a dentist made for him, adding a sharper pain to the already dull lingering pain of a rotting molar. When he and the others see the supervisors' pickup truck, they rise to their feet and rush toward it. Zamora squeezes his dust-covered body and bloody mouth into the narrow seat behind the driver.

Most of the other men jump into the bed of the pickup. "Go! Go! Let's get out of here!" they yell once everyone is on board. At the wheel, Avalos heads toward the surface. The Toyota Hilux pickup truck sags under the weight of two dozen men, "pushed together like bees in a hive," Carlos Mamani says. He's standing on the back bumper, wrapping his arms around the legs of the men standing in the truck bed. To the men in the cab, with the hood seeming to rise in the air, the pickup is like an overloaded aircraft straining to take flight. The dust once again gets too thick to see, so Mario Sepúlveda gets out of the cab and walks ahead with his flashlight, guiding the driver forward. Marching and driving this way into the dust cloud, they meet Raúl Bustos and the other three contract mechanics who were at the workshop, and the mechanics pile into the back of the pickup, too, all the while sharing the story of the explosions they felt from their perch on the edge of the cavern. Advancing deeper still into the dust, the men hear a mechanical rattling approaching: It's the personnel truck, with Franklin Lobos and Jorge Galleguillos. Sepúlveda shines his light in the faces of the older men and sees the blood-drained look of mortal fear. The older men recount the collapse they escaped, with Galleguillos insisting on having seen a butterfly moments before. Urzúa orders them to turn the truck around and head back uphill. When Lobos does so, most of the men leave the pickup and become his passengers. The ascent continues, but with each turn higher up in the spiral, past Levels 150 and 180, more debris appears in the roadway of the Ramp, as if they were getting closer to the scene of a battle that had been fought with stones. Up one curve, and then another, they go forward, until, after the eighth loop from the

Refuge, they approach Level 190. A few times they stop because the dust is too thick, and they wait five, ten, fifteen minutes for it to clear a bit. Finally, there are too many rocks to drive any farther, and all the men get out of the pickup truck and the personnel truck and walk on foot. Uphill, on a 10 percent grade, it's a climb that can quickly wear a man out, especially given the heat and humidity, but not for men being carried upward by their own adrenaline, trying to imagine the midday sunlight that awaits them at the end of this slow, oft-interrupted journey, if the Ramp has held together the way the mine's managers and owners have promised. They advance on foot another fifty yards or so, following the lights of their headlamps and flashlights through a gravelly cloud, until the beams strike an object that appears to be blocking the way forward. It's the gray surface of a stone slab, its size and shape not quite clear in the still swirling dust. They sit and wait in that cloud for several minutes, for the air to turn less gravelly. As it does, the full size of the obstacle before them becomes apparent.

The Ramp is blocked, from top to bottom, and all the way across, by a wall of rock.

To Luis Urzúa it looks "like the stone they put over Jesus's tomb." To others it is a curtain of rock, and to one miner a "guillotine" of stone. It's a flat, smooth sheet of bluish gray diorite, and it's dropped across the roadway of the Ramp in the same way trapdoors fall suddenly and theatrically in action-adventure movies. To Edison Peña it's the stone's newness that's most disturbing—it's clean, unsoiled by the soot and dust of the mine, as if created anew to trap them.

Only later will the men learn the awesome size of the obstacle before them, to be known in a Chilean government report as a "megabloque." A huge chunk of the mountain has fallen in a single piece. The miners are like men standing at the bottom of a granite cliff: The rock before them is about 550 feet tall. It weighs 700 million kilograms, or about 770,000 tons, twice the weight of the Empire State Building. Some of the men can already sense the enormity of the disaster. Mario Gómez believes, as do others, that the collapse originated up at Level 540, where a large crack had split the Ramp and leaked water several months earlier. That is where, at the insistence of Jorge Galleguillos and some of the older miners, the mine owners had placed mirrors in the crack to see if the rock was shifting. The mirrors had never broken, but in their gut, at this moment, Gómez and Galleguillos and the older men know the mine failed at that spot. By their quick (and largely correct) calculations, it's likely at least ten levels of the ramp have been wiped out.

"*Estamos cagados*," one miner says. Loose translation: We're fucked.

Alex Vega looks to his fellow miners to be the most desperate among them to leave. On an ordinary day, out in the sunshine, Alex has the muscled, melancholy handsomeness of a model in a cigarette ad, with longish sideburns and a well-defined brow. He's about five feet, three inches tall, and here before the stone he looks especially small, though his smallness is what gives him hope. He slithers onto his stomach and stares into a small opening beneath the gray stone blocking the way out—he might be the only guy who can fit into that space.

Like many people from the north, Alex Vega is a quiet homebody. He got his girlfriend, Jessica, pregnant at the age of fifteen and later married her, and they've been together for fifteen years since then. At the San José his nickname is "El Papi Ricky," after a soap opera character who, like Alex, is a father with a young daughter. Some years back Alex and Jessica took out a loan to buy an empty lot in Copiapó's Arturo Prat neighborhood, and they're slowly filling that lot with rooms, and building a low cinder-block wall around their property that's a symbol of their good fortune and hard work: It's just three feet tall now, and Alex has kept his well-paying job as a mechanic in the mine so that he can finish it, despite being warned by his father and two of his brothers (who used to work in the San José) about how dangerous the mine is. Alex wants to get back home, and the only path is this opening where the fallen block has blasted out the floor of the Ramp. He tells the men he thinks he can squeeze through.

"No," Urzúa says, and several other men say it's a crazy thing to do.

Vega insists, and finally Urzúa tells him, "Just be careful. We'll be out here listening, and if the rock starts to crack or move we'll tell you."

Vega squeezes his small frame into a crevice of jagged rock. "At that moment, I was feeling all this adrenaline," he later says. "I didn't think, or measure the risk." Not long afterward he'll think: *What a stupid thing I did*.

With his lamp in hand, he crawls twenty feet into the crack, until he can advance no farther.

"There's no way through," he announces after he crawls out.

For some of the older men and lifelong miners, the sight of the stone and Vega's words bring an overwhelming sense of finality. Many have been trapped in mines before, by small rock falls that a bulldozer can clear in a few hours, but this gray wall is something completely outside their experience. The flat stone is a vision of death, and it causes them to reflect, as they stand before it, on the world they've been separated from: the realm of the living, of families and fog-laden breezes, of homes and paternal obligations. All the things they are leaving unsettled in their lives begin to gather in their thoughts. Galleguillos thinks he'll never see his new grandson, and feels the tears running down his cheeks. Gómez realizes that, like his miner father, who died of silicosis, he's worked too long and pushed his luck until he had one accident too many: first two fingers, and now his life. He will die a miner. *This is as far as I'll ever go*, he thinks. *Hasta aquí llego*.

The men's confused silence is soon filled by the sound of the shift foreman counting. Raúl Villegas, who was driving one of the ore trucks, is missing, but Franklin Lobos and Jorge Galleguillos saw him on his way to the surface, and it seems likely that he got out. "Thirty, thirty-one, thirty-two . . ." Urzúa counts again, and again gets thirty-two, but all the men are shifting around, and he's not quite clear that this is the right count, because in the San José Mine the lists of workers always shift from one day to the next. Nothing in the mine is ever certain, though the supervisor is pretty sure of one thing: There's probably no way to escape, and no way for rescuers to reach them, in this haphazard collection of passageways cut into a mountain.

3

THE DINNER HOUR

There is nothing especially remarkable about the phone call the off-duty miner Pablo Ramirez gets at about 2:00 p.m. It's from Carlos Pinilla's secretary at the San Esteban Mining Company. "There's a problem in the mine," the secretary says. "A problem with the Ramp. Don Carlos says to get over here. And it looks like you'll only need a few operators for your shift." Ramirez is at home in Copiapó, enjoying his last few hours of rest before beginning work on the night shift at the San José Mine. When the shift currently inside the mine ends its workday, he will take the reins from Luis Urzúa as the supervisor of the night shift: but that's not supposed to be for another five hours or so. Judging from the secretary's matter-of-fact tone, there's been another spillover of rock from the cavern of the Pit, and getting the guys out will be a routine but arduous task involving a few machine operators clearing out rock from the Ramp. A day of production will be lost.

This is what Ramirez thinks as he drives to the mine. He's not worried for all the guys he knows inside. He knows about half of the men quite well: The shift foreman, Florencio Avalos, is one of his best friends, and Florencio's two sons call Pablo "*tío*," or uncle. Ramirez and Avalos are both about thirty, smart young guys with good careers ahead of them in mining, and at this moment Ramirez hasn't heard anything to disabuse him of the belief that he'll be able to sit down and have a beer with his friend fairly soon. At 4:30, Pablo arrives at the mine and sees something significantly more troubling than he expected. The mouth of the mine, its one and only entrance, is spewing dust. A bit of dust coming from the mine isn't that unusual, but Ramirez has never seen quite the billowing cloud he sees now, rising up from the cavelike entrance, "like a volcano," he later says. And then there are the noises: the explosions produced by rock falls, a moaning crash repeated again and again. But even this is not terribly unusual, because the mountain is constantly broadcasting noises from its innards to the outside world: Whenever a team of workers sets off a blast deep inside, for example, the sound usually follows the mine's passageways and rises to the surface. But neither the sound nor the spewing dirt stop. Past five, six o'clock, the dust is still making it impossible to enter the mine and go for the men inside. As the hours pass, mine workers and managers gather outside, looking a bit lost, with Carlos Pinilla standing there in his white helmet among them. Pinilla has tried twice before to go inside, reaching only as far as Level 440 before the cloud of dust became too thick to continue.

At about five o'clock, with the dusk of Southern Hemisphere winter fast approaching, Pinilla leads a team of men back inside: Pablo Ramirez and two other mine supervisors enter with him.

They descend in a pickup truck, making several switchback turns without incident, until, at Level 450, they see a two-inch-wide crack all across the floor of the Ramp. The "good" diorite of the only passageway into and out of the mine has fractured wider than Ramirez has ever seen before, and he will remember this as the moment he first grasped the seriousness of the accident. As they advance a bit farther, Ramirez follows the light of the pickup truck and expects to see, at any moment, the loose rock produced by a cave-in from the excavated cavern of the Pit. Instead, at Level 320, after driving 4.5 kilometers from the mine entrance (2.8 miles), the pickup comes to an unexpected obstacle, its headlights shining upon a flat, gray mass. The Ramp is blocked, top to bottom, by a single piece of rock. Ramirez thinks he's seen or prepared for every mining disaster possible, but never in his musing or his planning has he imagined anything like this. It feels as if someone had gone through with a knife and cut the mine in half. The men get out of the pickup and stand before it, a solid wall of rock that couldn't possibly be there.

"*Cagamos*," someone says, unknowingly repeating the same word uttered by the men trapped by this same rock 425 feet below. We're fucked. Ramirez, a man who takes pride in being able to tackle any problem a mine can give him, suddenly feels a sense of powerlessness, that there is nothing he or his boss can do to rescue the thirty-three men on the other side of this rock. The best remaining hope is the chimneys, but only a special police unit, a team with mountain climbing gear, will be able to enter them.

Standing before that solid stone in his white helmet, Pinilla looks stunned. He is the most powerful man in the daily working lives of the men of the San José Mine, the high-strung boss who hurriedly left behind his underlings just before this rock fell in his wake. But now he begins to weep. "He's usually a jerk, real macho when it comes to those things," Ramirez will say later. "But he started crying, right away."

"I thought, no, I knew for certain that someone had to be dead," Pinilla says later. At 1:45, the moment of the collapse, the personnel truck was supposed to be headed uphill, taking the men out for lunch; the contractor mechanics weren't even supposed to be in the mine at that hour, and more than likely they, too, were headed out of the mine for lunch. Pinilla imagines those men crushed in the massive failure of the Ramp and can't help but think: *I'm the bastard who sent them all down there.*

They drive back up to the surface, and find the owners, Marcelo Kemeny and Alejandro Bohn, waiting in white helmets. They have to call for help, there's no other way. This requires Bohn and Kemeny to get in their truck and drive down the hill away from the mine and toward the highway in search of a cell-phone signal—there's a phone line at the mine but for some reason they choose not to use it. At 7:22, more than five hours after the collapse, the owners of the San José Mine call the authorities for the first time.

The call reaches the local fire department, then the offices of the National Geology and Mining Service, and eventually the disaster office of Chile's Ministry of the Interior, which supervises all of Chile's police and security forces. An hour later, six men from the Chilean police's Special Operations Group (GOPE, in Spanish) arrive at the San José Mine with climbing gear. They enter the mine in a pickup truck and are on the Ramp, passing Level 450, when they get a flat tire while driving over the new crack in the roadway. Following behind them in another pickup, Carlos Pinilla and Pablo Ramirez see the crack has doubled in size.

If the police unit were to know that the crack is new, and how much it's grown in the past two hours, they might realize how unstable the mountain is and halt their rescue effort. So in Ramirez's account of the moment, as the police rescuers get out to quickly change the tire, Pinilla looks over to Ramirez and places a finger against his lips. Ramirez understands and doesn't say a word about this frightening, growing fissure in the mountain, a warning sign of another collapse that will surely follow.

As the hours pass with thirty-three men trapped behind a cloud of dust and a curtain of stone, the administrators of the San Esteban Mining Company put off calling the families of the men. "In certain mines, the first instinct is to just try and hide these things as much as possible," one Chilean official later says. At 3:00 p.m., they might have called and said: There's been a collapse, it's going to take a while to get them out, don't expect them for dinner. At 7:00 p.m., they might have said: It's looking more serious than we expected, and we don't know when exactly we'll get them out, but as far as we know they're safe. Instead, for more than eight hours after the accident and deep into a Thursday night rumbling toward a Friday morning, the company's representatives say nothing, and word of the accident first reaches wives, mothers, fathers, daughters, and sons in Copiapó and other cities and towns across the narrow spine of Chile via the vague, alarming, and often inaccurate bulletins of various radio and television stations.

One of the few loved ones to hear quickly from someone with more or less direct knowledge of what's going at the mine is a woman not listed on any mining company document, and with no legal claim to any private information: Yonni Barrios's big, rosy-cheeked, and incessantly happy mistress, Susana Valenzuela.

Susana's brother-in-law works in a mine in Punta del Cobre, just outside of Copiapó, and learns of the accident as word goes out to the mining community that men might be needed for a rescue. That brother-in-law calls his wife, Susana's sister, who at 7:00 p.m. shows up at the door to Yonni and Susana's home in Copiapó and asks, "Is Yonni here?"

"No, he's not," Susana says.

The sister puts Susana in touch with her husband. "There was a cave-in at the mine at two in the afternoon and *los niños* are buried alive," he says over the phone. "There's no escape."

"That's what he told me: 'Buried alive.' 'No escape,'" Susana says later. "People who work in the mines know what it means to say those things. So I got desperate."

With her sister, Susana heads to the local station of Chile's national police force, the famously erect, incorruptible, and efficient Carabineros. But the police haven't heard anything either. In the short time Susana and her sister are there, however, word reaches the station, and they watch as police vehicles start to head out to the San José. Go to the hospital, the Carabineros tell them, but first Susana and her sister race back to their neighborhood, to tell Marta, Yonni's wife.

Marta is a much smaller, older, more severe and serious woman than Susana. The women have known each other for several years, as Yonni has bounced back and forth between their two homes. Susana first met Yonni in his wife's house: She mentioned to Marta that she needed someone to build some furniture. "This ugly old man I live with, my husband, made that over there," Marta said. "I'm bored of him." Yonni then emerged from the room where his wife had him "prisoner," or so Susana tells it. Marta explains that she's endured his philandering for years. Susana thinks: *This guy isn't ugly at all.* In his sad, lonely smile there is a hint of sly cunning, the come-hither look of a wounded man who wants to open his soul to you, right now, as soon as you can slip away and be alone with him. "I brought him over here, and I liked him," Susana says, "and I made him a little lunch, and then we made a little love afterwards." He never did make the furniture. Now this history is a light and comical prologue to the tragedy of that same philandering man buried and lost in the San José Mine. When she learns of the accident from her husband's mistress, Marta responds in a rather bloodless tone: "This is as far as you go with him. Now I'm in charge. Go get me the marriage book." The *libreta de matrimonio* is a kind of passbook, signed by the official who performs a civil marriage ceremony, and it's used to apply for a variety of government services, and also to gain access to places where only a spouse will be admitted (for example, a hospital room or a coroner's office). This document is in Yonni's possession, at Susana's home.

Susana obediently retrieves her boyfriend's marriage book, and together, she and his wife head off to the hospital.

Carmen Berríos is in Copiapó, on the bus, and the driver is playing fast-tempoed Mexican music on the radio and forcing all his passengers to listen to it. She's spent the day with her father and is headed back home, to make dinner for her husband, Luis Urzúa, and their two children at 9:30. Suddenly, the swirling accordions give way to an announcer's voice. "Extra, extra, extra!" the announcer says, in a radio voice that can't help but attach a note of cheerful anticipation to the dispensation of a news flash. "Tragedy in the San José Mine! A collapse in the mine!" The flash dispenses little more than the bare skeleton of a story before the Mexican music returns. For Carmen, learning that she might have lost her husband while that folksy music plays is a juxtaposition that's odd, cruel, and unforgettable.

"Driver, what did he say?" she asks. Because, to be honest, Carmen's not entirely sure exactly where Luis works. He switched jobs a few months back, and she's never been to his new mine, and she can't be sure the mine in the radio bulletin is the one where Luis works. For a moment her small doubt becomes a source of hope. "Can you change it to another station?" she asks the driver. "Maybe there's more information . . ."

"That's all there is," the driver says. "It's an extra. Maybe they'll have more later."

She gets home and hears more bulletins on the radio: There are wounded miners, and dead miners, the radio says, and it's now undeniably clear it's Luis's mine. "The whole nightmare fell on top of us, because we didn't know if it was true, or not." The clock in the living room passes 9:30 and Luis doesn't come home. Her two children, a son and a daughter both in college, are quietly studying and not especially aware that dinner is late. When the clock reaches 10:30 she calls them together and says: "We have to have a family meeting. The radio is saying there was an accident at the mine where your father works." She turns on the radio for them to hear, and there are reports of injured men being taken to the San José del Carmen Hospital in Copiapó, but Carmen decides she has to go to the mine, to get confirmation that he was even there when the mine collapsed. Her daughter's friend drives her there in a pickup truck, using a GPS, because she's never been there before and doesn't know the directions.

Carmen Berríos rides through an eerie night landscape filled with the shadows of round and rocky mountains, remembering a recent dream she had of Luis trapped underground: In that dream, he was rescued and rose to the surface in a bus. The pickup turns off the highway and onto the short spur leading to the mine, a mountain with a brightening cluster of lights, until they reach the front gate and the guard shack there, and she steps out into a night of dry and bitter cold. It's nearly midnight.

At her home in a middle-middle-class neighborhood of Copiapó, Mónica Avalos, the wife of the foreman, Florencio Avalos, is sewing. She's got one of her sixteen-year-old son's sweaters and is taking it in a bit, while also cooking a soup for Florencio. Her husband is a big soup eater, and Mónica is making a concoction of chopped beef she'll remember as being especially hearty, the way Florencio likes it. The smell of this meal is filling the living room and the small, attached dining room where she and Florencio and their two sons gather every night for dinner. Mónica is not watching television or listening to the radio, because she likes the silence in the house, and her sons are off in their rooms. Her main company at this moment is the big clock in the living room, which is marching toward the time Florencio usually comes home and she serves dinner: 9:30 p.m., as is the South American custom.

The phone rings with a call from her sister. "Look, I don't want to worry you, but there was a collapse in the mine. A really big collapse. In the mine where Florencio works. The San José Mine. Is Florencio there yet?"

"No, but he's going to be here any minute now." This call is followed by the arrival of 9:30 and several long, long minutes with Florencio's chair at the dining room table still empty. Suddenly, Mónica can't even remember the name of the mine where Florencio works. Was it the San José? She remembers him saying he worked in a mine named for another saint: San Antonio. Anthony, not Joseph, that's what she's thinking.

She's walking up and down the stairs, in a kind of manic trance, looking at that clock push further away from 9:30. Her seven-year-old son, who's alone in his room watching television, and who's always talked to his father about work and knows perfectly well what the name of the mine is, suddenly comes running into the living room and yells: "Mommy, my father is dead! My father is dead!"

"No! No!" his mother answers back. "Where did you get that from?"

"Don't be a liar!" the boy yells. "It's on the news!"

Upstairs, in her younger son's room before the television set, Mónica faints. Her older son, César Alexis, "Ale," comes to revive her, and to be steady and to play the role of father suddenly. Ale is sixteen years old, the same age Florencio and Mónica were when she got pregnant with him, and suddenly he is calm and strong, as if he were channeling Florencio somehow.

"*Cálmate*," Ale says to his mother. "*Cálmate*." They decide to go to Pablo Ramirez's house, because Pablo works in the same mine, and if anyone knows the truth and can be trusted to tell them, it's Pablo. When Mónica and her two sons arrive at Pablo's house and knock at the door, they are unaware that Pablo is, at that moment, entering the collapsed mine in search of Florencio and the other thirty-two men with him. No one answers for fifteen minutes, until finally Pablo's wife comes to the door and says: "Pablo's not here. He went to the mine. Because there's been an accident." Mónica calls another of Florencio's friends, Isaías, and they drive to the mine together, but get lost on the roads outside of town. When she arrives, she takes note of the emptiness of the place. She sees men in helmets and uniforms walking back and forth with unknown purpose before the mine entrance. It feels like she's the first and only woman there.

In Talcahuano, Carola Bustos, who survived the earthquake and tsunami with her husband, decides she can't bear to tell her two young children what's happened to their father. They will hear her voice breaking and see her weeping, and the sight of their stricken mother will wound them. To spare her children that hurt, she leaves them in the care of her parents, in the physical and emotional safety of their home, and she slips away for a northward flight without saying goodbye. Carola leaves it to her parents to explain her absence to their grandchildren: "Mommy's going back to Santiago to look for work, and she'll be back soon."

Some seventeen hours after the collapse, the phone rings at the home of Mario "Perri" Sepúlveda in Santiago. Elvira, who goes by the nickname "Cati," takes the call from a friend just after 7:00 a.m.

"Cati, there was an accident in a mine and it looks like Mario is there," says a friend.

For a moment, it seems like a joke. "No," Elvira says. It seems impossible that a friend from Santiago would know anything about what Mario was doing hundreds of kilometers to the north.

"I'm not bullshitting you," the friend says. *No te estoy huevando.* "Turn on Channel Seven."

Elvira turns on the television and sees the report from Copiapó. After a few moments, Mario is on the screen, a clean-shaven and not especially happy man of forty, staring into the camera with red, flashbulb eyes in the unflattering picture on his mining company identity card. His full name is there as a caption: MARIO SEPÚLVEDA ESPINACE. The news report fills out some details. The collapse took place yesterday afternoon, they're buried several hundred meters underground, all communication has been lost. After she digests the gravity of what's happened, when she has time to think about exactly what her husband must be suffering, she wonders: *How is that crazy man going to survive cooped up like that? He needs to be moving around. He won't be able to take it.*

As to the accident itself, Elvira is not surprised: Mario had more or less predicted it. When he went to work, Mario often reminded her about the insurance and social security she would be entitled to if anything happened to him in the mine. He spoke so often and angrily about the San José being on the verge of collapse, his worries invaded the dreams of his eighteen-year-old daughter, Scarlette. One day, several months earlier, Scarlette had a nightmare in which she learned her father had been killed in the mine. She woke up screaming "My father is dead!" and could not be convinced that it was just a dream. She was crying and trembling, and her mother was forced to take her to the hospital, until Mario emerged from his shift and called home and said: "Scarlette, it's me, your father! I'm alive! I'm fine! Nothing happened to me. I was just at work . . ."

With Mario now buried, Elvira can't help but think of Scarlette's dreams as a kind of a prophecy no one chose to heed. She wonders: *How am I going to explain to our son that his father is trapped and there's nothing we can do?* Francisco is thirteen years old but is small for his age, and always has been. He was born after a pregnancy of just five months, weighed a mere 2.4 pounds at birth, and spent the first sixty-nine days of his life in an incubator. They are very close, father and son, their bond having been forged during those ten weeks when a powerless Mario was forced to endure the sight of his baby boy in a box, with his impossibly tiny limbs and closed eyes, being fed by tubes, seemingly fighting for his life with clenched fists as small as rosebuds. Mario has filled up the rest of the boy's days with as much love as he can give him. He's become the boy's personal cheerleader, comedian, and preacher, leading him on a series of outdoor adventures and pontificating, always, on the wonders and idiosyncrasies of electrically operated machines, and on the common sense and sensitivity it takes to care for both horses and dogs, and also on the Sepúlveda family's rural roots as *huasos*, the poncho-wearing, horse-riding Chilean equivalent of cowboys. She has watched father and son ride horses and kick around a soccer ball together, and sit before a television screen again and again, enraptured by a movie about fatherhood, loyalty, and warfare that's Mario's favorite: *Braveheart*. Mario is a big Mel Gibson fan "because me and my son, we aren't tall, and neither is Mel Gibson." Gibson's Academy Award–winning film is titled *Corazón Valiente* in Spanish. "Your heart is free," the movie says. "Have the courage to follow it."

Mario has told his son, "I am your Corazón Valiente," and now that Chilean-miner Braveheart is on the television. First in his employee photograph and then, most improbably, in a video in which he's talking to the camera, laughing, being Perri.

The only images we have of the miners are these recorded by Mario Sepúlveda, the television says. Mario loves to film things, and there he is, narrating a description of the bunk beds in the house where he stays with his fellow out-of-town workers during the seven days he's in Copiapó.

Elvira travels to the airport for the flight to a city she's never visited. Later that afternoon she's over the southern edge of the Atacama Desert with her weeping son, who can't stop saying how much he misses his father. And also with a daughter who was once hospitalized for believing a dream that seemed like madness, but which has now taken form in the waking world, in images and words broadcast and repeated on all the televisions and radios around her as she steps off the plane, into the light of a desert winter: Mario Sepúlveda Espinace, father of two. Feared lost in the San José Mine.

4

"I'M ALWAYS HUNGRY"

The trapped men eventually begin to turn their backs on the curtain of stone that separates them from the surface. They split into two groups. The first, smaller group decides to search in the mine's matrix of intersecting tunnels and excavated holes for a passageway to the surface. About eight men in all, they head for one of the cylindrical chimney shafts bored into the stone between different levels of the Ramp. The main purpose of these chimneys is to allow air, water, and electricity to flow into the mine, but they were also supposed to have been fitted with ladders to provide an escape route between each level. In theory it should be possible to climb about ten such passages and make it past the collapsed section of the mine, but in practice only a few of the chimneys have ladders, and *los niños* don't have much faith that they'll be able to find a path upward. Still, they start off for the nearest chimney opening, a short walk downhill to Level 180.

A second, much larger group of about two dozen men heads back to the Refuge to wait, walking back downhill toward the personnel truck. As the two groups split up, Florencio Avalos, the shift's foreman and second-in-command, takes Yonni Barrios aside. "Down in the Refuge, take care of those two boxes with provisions," he tells Barrios. "Don't let *los niños* eat them yet, because we may be trapped for days." Avalos says this sotto voce, because he doesn't want to spread a panic among the men. He chooses to tell this to Barrios, because Yonni is among the oldest and most experienced in the group—and also because Yonni is the kind of guy who will follow any order you give him. "Yonni, I'm trusting you, I want you to take care of that cabinet for me. Don't let anyone open it until we get back, please."

Carlos Mamani, the Bolivian immigrant, joins the line of walking men and realizes how much he needs that lamp he left in his locker on the surface. The supervisor testing him out on the Volvo L120 loader had said it wasn't urgent to have one and he could pick it up during the lunch break. Now Mamani is going to have to come to terms with his natural, human fear of the dark, because without a lamp, darkness is his constant companion. As the line of marching men spreads out on the walk downhill, he finds himself inside a patch of blackness, walking uncertainly between the silhouettes of the boulders scattered across the roadway, looking up to follow the cones and beams of incandescent light shining from the helmets of other miners, until they finally reach the personnel truck and he jumps in the back for the ride down to the Refuge.

Upon their arrival, and after a quick exploration of the nearby area, the men take note of the fact that all the connections to the surface have been cut: the electricity, the intercom system, the flow of water and compressed air. Despite this very bad sign, there are still some among the trapped men—especially those with less experience in underground mining—who think they'll be rescued within a day, or perhaps in a few hours. Those first few hours begin to pass, slowly, punctuated by a rumbling stomach or two (they have just missed lunch, after all), and by the continual thunder of rock falling somewhere in the dark spaces beyond the weak, warm light of their headlamps, many of which the men now begin to turn off, to save their batteries. When you're hungry, waiting to eat is an ordeal, and even more so when you're sitting in a room with two locked cabinets filled with food, a stock of calories guarded by one rather timorous, middle-aged miner. Yonni Barrios, the man who can't stand up to either of the women in his life, now has the responsibility of trying to keep two dozen hungry men from eating the emergency supplies for dinner.

The small escape expedition begins with one miner driving a jumbo lifter with a long arm and a platform to the chimney opening. This piece of equipment is usually employed to lift the men who fortify the mine's ceilings and who bore holes into the rock to place explosive charges. Mario Sepúlveda climbs into the basket and is lifted up into a hole in the ceiling. Raúl Bustos, the mechanic from the port city of Talcahuano, follows after him. For Luis Urzúa, the shift supervisor, scaling the chimney is at once dangerous and useless, and he has made a halfhearted effort to keep the men from trying. "Climbing up that chimney wasn't going to work. None of those guys were thinking about safety. And you'll notice that the guys who went up there first weren't really miners," he later says, referring to Sepúlveda and Bustos, neither of whom is from a mining family. But Urzúa is quickly losing his authority over the men, who are determined to do something to try to save themselves. Raising his head into the two-meter-diameter hole, Sepúlveda is surprised to see a ladder, built from pieces of rebar driven into the rock. He begins to climb, hopeful that he'll find a way out. After a minute or so, he realizes he's really too overweight to do this, but he keeps on going. It's a hundred feet or so to the next level, the hole rising on an incline. The taste of dust and vehicle exhaust makes it hard to breathe. Behind them are two other men, Florencio Avalos and Carlos Barrios, a twenty-seven-year-old whose girlfriend has not yet told him she's pregnant with his child. The walls are slippery with humidity, and soon the four men are covered with sweat, too. Halfway up, one of the rebar rungs breaks off as Sepúlveda grabs it, and the metal strikes him in the front teeth, sending a rush of blood into his mouth. He shakes his head in pain, but keeps climbing.

Raúl Bustos can hear Sepúlveda breathing heavily above him, and climbs after him until he knocks loose a huge slab of rock—Bustos can feel the slab moving, and he keeps it from falling by pinning it against the chimney wall with his shoulder. He yells to the two men below him: "Go down! Go down!" He pushes his hands, feet, and head against the stone as Avalos and Barrios scurry to the bottom, and when they've reached safety he groans and lets it fall. It hits the sides of the chimney several times as it crashes downward, finally landing on the Ramp below, hurting no one.

Up above, Sepúlveda reaches the top of the chimney and sweeps the beam of his flashlight across the blackness: more rocks are scattered on the floor of the Ramp here than on the level below. He stands up, and when Bustos finally reaches the top they begin to walk, following the Ramp upward, with the faint hope that after the next curve in the spiral the route to the top will be open. Instead their light beams strike a shiny, smooth surface: a curtain of diorite identical to the one blocking the Ramp down below. Mario can feel the hope draining from his body, and in its absence he is left with a clear, cold vision of what is happening to him. He is trapped underground, suddenly and unexpectedly close to death, but still in control of his fate. "At that moment I put death in my head and decided I would live with it," he says later. They walk downhill, past the chimney opening they've just scaled, and go around another curve to find the same gray "guillotine" blocking their path again. When they look for the next chimney opening, the one that might take them up to a higher level, their flashlights reveal that in this one there is no ladder at all, but instead just a cable dangling inside.

"This way isn't going to work," Sepúlveda tells Bustos. "What are we going to tell *los niños*?"

"It's hard," Bustos says. "But let's tell them the truth."

Down at the bottom of the chimney, Sepúlveda and Bustos deliver the news to the small group waiting there. The Ramp is blocked on other levels, too. There's no way out.

Collectively the men around them think: *Now what?* Their eyes settle on the man in charge, Urzúa. The shift supervisor says nothing. He's usually the most relaxed and upbeat of bosses, but at this moment he looks drained and defeated. His green eyes drift nervously: They look away from his underlings, who get the sense he wishes he'd never been appointed shift supervisor, and that he could just fade into the background and be one of the guys. Many of the miners, especially the oldest ones, come to believe that "the shift supervisor disappeared on us" in these first hours following the accident. It's especially grating to the older men because they believe mining hierarchies have a purpose, especially in a crisis. To be trapped in an underground mine is like being on a ship in a storm: The captain has to take charge. Instead, a few hours later, according to several miners, Urzúa will slip away from the group and go off to his pickup truck and lie down in the front seat, alone.

Urzúa explains his actions this way. He's a trained topographer, and thus carries a map of the mine in his head; that map tells him there's nothing to be done. "My problem as the *jefe de turno* was I knew we were screwed," he says later. "Florencio and I knew this, but we couldn't tell that to anyone. You practically had your hands tied from doing anything. And you start imagining things that are part of the reality of mining." The reality of mining is that sometimes men are buried alive and eventually die of starvation, with only a small chance that their bodies will be recovered later. An even harsher reality of Chilean mining is that after six or seven days, if the rescuers don't find you, they give up. Urzúa knows this, and also knows he'd sow a panic if he were to speak these hard truths. That was the most important lesson he learned from a recent course he took, "Responding to Critical Mining Situations": Stay calm. The only thing to do is wait—most likely for some sort of attempt to reach them by drilling from the top, which is really their only hope. But how can he inspire men to wait, to do nothing? He cannot. That's how Urzúa sees things. Already he's doing the math in his head and realizes that a rescue from the top will require a feat of drilling never before seen in the annals of mining. He'd like to say something to give the men hope, but he cannot, for the simple reason that he refuses to lie. Instead he says nothing. If his job is to get the men out safely, the only thing he can do, really, is to keep them from doing anything stupid or rash: like, for example, trying to climb out through the caverns of loose, constantly falling rock in the Pit. Waiting is the better option. *Just keep calm*, he tells himself, working hard not to show his underlings the desperation he's feeling.

Later, in a moment of quiet, when they're waiting for a rescue that may or may not reach them, Sepúlveda will tell the shift supervisor how much he admires his steadiness under pressure. "Lucho Urzúa has a problem with speaking, with a lack of passion," Sepúlveda later says. "But he's very smart and wise." Urzúa has decided that when he reaches the Refuge and has all the men of the morning A shift before him, he's going to announce that he's no longer their boss. They're all stuck together, and they should make decisions together, he thinks.

Sepúlveda's response to the grim situation in which the men find themselves is entirely different. He summarizes his attitude at this moment with a vulgar Chilean phrase: *tomar la hueva*, which can be loosely translated as "gripping the bull by the balls." His life, up to and including this moment, has been one struggle for survival after another, and in a certain sense fighting to stay alive is when he most feels like himself.

Mario's mother died delivering him, and Mario believes he was born fighting. He grew up in the southern city of Parral (the poet Pablo Neruda's birthplace) with ten brothers and sisters, the son of a peasant father who scarred his sons with his drinking and his discipline, directing much of his anger at the self-described "hyperkinetic" Mario. "When I was twelve years old I was so hyper I frightened people. None of my relatives wanted to take care of me." Today such a boy might be given an ADHD diagnosis and a prescription to deal with his expansive energy. Back then, his father tried to beat it out of him. Against the routine administration of doses of swinging leather, there was the calm and dignified presence of Mario's grandfather, a *huaso* who imparted the country ethic of hard work, respect, and personal integrity. Mario started working in and around Parral at thirteen, then left at nineteen for Santiago. He had to leave behind his younger brothers—born to his stepmother— unprotected against their father's rage. "I will be back for you," he told José, David, Pablo, and Fabián, and he kept his promise. He found a job as a *barrendero*, or sweeper, with a Santiago company, and settled in a barrio where his neighbors came to know him as a brash street pugilist whose ferociousness could quickly give way to a charming provincial formality. During the day he swept floors and at night he'd dress in suits and flared pants inspired by John Travolta and *Saturday Night Fever*. He fell in love with a young woman named Elvira who favored the glam-rock-inspired stylings of early Madonna and Sheila E. She remembers the young Mario as a man of shifting moods, quick to anger, but quick to forgive. He was also sentimental and open about his feelings, a rare quality for hot-blooded men. In sum, he was possessed of a rough-hewn nobility. One Chilean expression describes the way he lived his life then and now, and Mario uses it a lot: *tirar para arriba*, which literally means "to toss upward," but which is most often translated as "to overcome adversity." Moving ever upward, Mario added new skills to his résumé, including the operation of heavy machinery, and never failed to provide for Elvira and their two children, even if it meant going to opposite ends of Chile in search of work: from the Atacama Desert in the north to the southern port of Puerto Montt, the gateway to Chilean Patagonia and the Strait of Magellan.

In his Santiago neighborhood, Mario's short-cropped hair has earned him the nickname "Kiwi"; but that same hair, over the intense stare of his brown eyes, can give him the menacing appearance of a man on the brink. To his fellow miners in the San José, "Perri" Sepúlveda often looks like a man possessed—even during an ordinary workday. Now, despite his lack of standing in the mining-shift hierarchy, the possessed Perri is going to try to take control of his own fate and the men around him with his optimism, his raspy voice, and his stray-dog sense of survival and loyalty to the people close to him.

"*Yo lo único que hago es vivir,*" he says. The only thing I do is live.

He and the other members of the failed escape expedition head back down to the Refuge. Unbeknownst to them, a pathetic drama is unfolding there. The hour when the A shift should have ended has long since passed, and some of the twenty-five men down there are very hungry.

"We have to break it open!" one of the men calls out. "We're hungry."

"*¡Tenemos hambre!*"

The cabinet around which they are gathered is supposed to contain enough food to keep twenty-five men alive for forty-eight hours in the event of an emergency. Many of the men haven't eaten since dinner the night before, to avoid the vomiting caused by working underground in intense heat, humidity, and dust-filled air. At about this time, they would be home, sitting at dinner tables while wives, girlfriends, and mothers serve them. A few say they should wait until the foreman and the others come back. In his low, gentle voice, Yonni Barrios says as much: "We need to wait, because we don't know how long we'll be down here." Some of the men have started to strip off their overalls against the heat. Half-undressed and sweating profusely, they begin to examine the sealed box.

The mild-mannered Yonni can see that stopping this hungry group of men will be impossible. "There were just too many of them," he later says, though none of his fellow workers remembers Yonni saying much of anything at this moment, as Víctor Zamora and another man try taking a screwdriver to the box's hinges, and also to the three metal strips wrapped around the boxes, sealing them like aluminum chastity bands.

Zamora seems most determined to eat: Getting hit in the mouth and losing teeth during the collapse has only worsened his mood. "*¡Siempre ando con hambre!*" he shouts as he tries to open the cabinet. I'm always hungry.

For many reasons, it is not surprising that Zamora is leading the assault on the food supply. He is a perpetual outsider, with tattoos on his forearms that serve as announcements of his status. One is a portrait of Ernesto "Che" Guevara, the militant saint of the Latin American poor, and another proclaims a single, meaningful word: "ARICA." Zamora was born and raised in that city, which Chile took from Peru in the nineteenth-century War of the Pacific, and he possesses coloring and features that some might call Incan. A few news reports have already incorrectly identified him as Peruvian. One of his fellow miners, who doesn't like or trust Zamora, disparagingly calls him "*el peruanito,*" the little Peruvian guy, even though he knows Zamora is as Chilean as he is.

Zamora's father died when he was eight months old, and his mother abandoned him, "because she preferred to be with her new boyfriend instead." He was raised by his mother's sister, but at the age of nine she sent him to an Arica home for street children, where he lived off and on until he was sixteen. "From the time I was small I wanted to have a family, but I never had one. I could see that everything was for everyone else and that I would get what was left behind, to live on the streets, sleep under bridges, die of hunger." This hard, lonely boyhood gave Víctor great cunning—and also an appreciation for the power of love. He labored his way out of poverty and into a confident adulthood where his growing sense of self-worth was defined by a slow ascent through the strata of unskilled jobs available to him (picking grapes out in the sun, raising up beams at construction sites), and also by his ability to hold on to the affections of Jessica Segovia, the mother of his son, Arturo. He met Jessica at a neighborhood fiesta when he was a wandering young man, "*gitiando,*" he says, living like a gypsy. At the San José, he's a fortifier, drilling steel anchors into stone walls. Víctor's temperament is suited to this work, which is exhausting enough to take a bit of the monster out of him each time he goes in, making him less likely to lose his patience with Jessica when he gets home.

Home is in Tierra Amarilla, just outside Copiapó, in a few rooms with low ceilings and pink plaster walls with a crack or two, where a couch and table are squeezed into a small dining/living room. Víctor will raise his voice to his family and fill that small room with the ugly thoughts, suddenly liberated and expressed, that can enter a man's head when he feels trapped rather than nurtured by family and its obligations. He is explosively angry, and then hates himself for being this way. He fights with his brother, too, who often wounds Víctor with the most cutting words he can speak: "You're not really my brother and my mother isn't even your mother," which is true, of course, because Víctor's "mother" is really his aunt and his "brother" is really his cousin. With his poetic sensibility, Víctor boils down his personality to a few well-selected words: His personality is "*polvorilla*," resembling the combustible qualities of gunpowder; at home, he's inflicted "*descontroles*" on his family, moments when he lost control of himself and his emotion. "A family isn't just all happiness," he says.

Víctor Zamora loves his mining family as much as he loves his real family, but at this moment, as he tries to open the box with the emergency provisions, he isn't really thinking about how his actions might hurt his underground brothers. Instead, when the screwdriver doesn't work, Zamora does something that he might have done when he was a young man trying to survive on the streets of Arica: He retrieves a bolt cutter (the same tool he was using earlier to cut rebar for his fortifying crew), steps forward, and snaps the metal bands on the box containing the emergency supplies.

He's about to break the locks, too, when the driver Franklin Lobos steps forward and says, "Wait, I have the key."

Lobos is taller and bulkier than just about anyone else trapped in the mine. His height is a reminder of the athleticism that intimidated people on the soccer field, and sometimes he uses his size to assert himself, suddenly dropping his unassuming air to give voice to some grievance, or to express annoyance at the fucked-up nature of this workplace. But at this moment Lobos decides that giving in to the hungry men is his only recourse. "I wasn't going to fight five or six of them. In the state that we were in, fighting didn't make any sense." This same idea will soon be repeated, often, in the thoughts of many of the trapped men as more conflicts and disagreements simmer between them: I really want to punch this idiot, this *huevón*, but I don't want to be stuck underground taking care of a miner with a broken jaw, or a bleeding wound either.

Lobos opens the cabinet and the rebellious miners' main object of desire is revealed: packages of cookies with the brand name Cartoons. They're really children's snacks, chocolate- and lemon-flavored sandwich creams, the kind you can split in half easily, several dozen packages in all. "It didn't seem like there were that many," Zamora will remember. Outside, in the surface world, you can buy these packages for 100 Chilean pesos each, or less than a quarter. There are four cookies in each package—several of which are quickly dispensed to those who will take them, though many miners refuse. Zamora will say later that he didn't think much about what he was doing. "I was just hungry. It was time to eat. I didn't give it much importance."

They open some of the milk boxes they find, too. About ten of the two dozen men present in the Refuge partake of this food, getting one or more precious packages of cookies each, sharing two liters of milk.

"It was the northerners that did it," one of the southerners later says. "They only thought of saving themselves at that moment. Fresh guys. They wanted everything. They never thought that we'd be trapped so long."

Later, one of the miners will remember listening to the looters of the food box eat in the darkness, sitting in a corner of the Refuge with their headlamps off, as if they were ashamed of their own hunger, and yet unable to keep the crackle of crumpling plastic and the moist crunch of their chewing from filling the small space, to be heard by men who did not take any food at all.

When Luis Urzúa and the party of men involved in the failed escape expedition arrive at the Refuge, they find a scene of disarray. They discover the unlocked cabinet, its severed aluminum bands. Gathering the discarded packages of cookies, they count ten. "With what you guys just ate, we all could have survived three days down here," Florencio Avalos says. "Well, whoever ate that food, let them get something out of it . . . May it serve them well."

The mood changes suddenly, as the members of the escape expedition reveal the truth they learned higher up in the mine: They're trapped and there will be no easy rescue or escape. They speak with a grave tone

of focused urgency that catches many of the men in the Refuge by surprise. "What are you guys doing?" Mario Sepúlveda says, with the raspy,
high-pitched paternalistic lament he might use for one of his dogs. "Don't
you realize we might be down here for days? Or weeks?"

No one immediately confesses to the crime of looting the provisions. Nor do any of the members of the failed escape expedition demand
to know who was responsible, and in the confusion of the moment, a
few men will not learn for many days what exactly transpired with the
food supply. The fact of the taking will simply sit for days to come in
the conscience of the men who did it. Víctor Zamora, recognized by
many as the chief culprit, studies the faces of his friends and companions
and understands, for the first time, the severity of what's happened in that
white-tiled room with the box that used to be sealed. He says nothing,
and won't say anything for days about what he's done.

Now Mario Sepúlveda and Raúl Bustos begin to give details of their
climb to the top, with Sepúlveda getting on his knees in the dirt to draw
a diagram of the blocked ramp, and the chimney without the ladder. He
speaks to them with a common term of endearment among men:
chiquillos, or "kids." "In other words, *chiquillos*, even if we're superopti-
mistic about things, the best you can say is we're in a load of shit. The
only thing we can do is to be strong, superdisciplined, and united."

In the silence that follows Sepúlveda's assessment, Luis Urzúa
steps forward to make an announcement. Given the circumstances, he
says, "We are all equal now. I take off my white helmet. There are no
bosses and employees." He is surrendering, in effect, his responsibilities
as a shift supervisor, as *jefe de turno*. A few minutes earlier on the walk
down from the chimney, Urzúa had told the members of the escape
expedition that he was going to do this, and now he's gone ahead and
done it, even though they told him not to. "We have to decide, together,
what we'll do," Urzúa says. What he wants to communicate is their need
to stand together and stand united, "one for all and all for one," though
what some take from this small speech and his low-key demeanor is
meekness in the face of a challenge, and the sense that the man who's
supposed to be in charge isn't.

"Sometimes Luis Urzúa says things without thinking," Raúl Bustos
later says. Bustos feels a kind of suppressed anarchy lingering there in

the cavernous space underground. Five months earlier, he saw his hometown of Talcahuano descend into anarchy following a tsunami and earthquake, and he was nearly mugged outside a pharmacy that was being looted. Like a natural disaster, the collapse might cause the daily order and hierarchies of the mine to come apart. Bustos can see the possibility that the strongest and most desperate men in the cavern will take advantage of the weaker men. The logic of the street could take over the mine. There are, after all, a few among them who have spent a bit of time in jail, for fights in bars, that sort of thing, and each of those men is a potential "alpha dog," he thinks. "At any moment, they could have turned on *el jefe de turno* if we didn't back him up."

Urzúa speaks his piece, but it leaves a kind of void, so others from the escape expedition try to fill it: Mario Sepúlveda and the foreman Florencio Avalos, and Juan Carlos Aguilar, the supervisor of the contract mechanics crew. *El jefe de turno* is correct, they all say. We have to stick together. Aguilar speaks with a voice that is at once authoritative and informed. The situation is not good, he says, but there are things they can do to prepare. Number one, they have to take care of all the water that's down there with them, because the water they used to keep the machines and the mine running can keep them alive, too. It's obvious they'll have to ration the food, too, eating as little as possible every day to make it stretch out as long as possible, and the only question is how to do it.

Sepúlveda helps lead a tally of what is (and was) inside the emergency cabinet: 1 can of salmon, 1 can of peaches, 1 can of peas, 18 cans of tuna, 24 liters of condensed milk (8 of which are spoiled), 93 packages of cookies (including those that have just been eaten), and some expired medicines. There are also, incongruously, 240 plastic spoons and forks, and a mere 10 bottles of water, which serve as further proof of the mine owners' thoughtlessness. The men will not die of dehydration, however, because there's several thousand liters of industrial water in the big tanks nearby that's used to keep the engines cool, and even though it's probably tainted with small amounts of oil, it will likely be drinkable. But they have to divide those cookies and cans of tuna among them all: If each man eats one or two cookies and a spoonful of tuna each day, the provisions might stretch out a week. They put all the food back into the cabinet, and lock it again. Urzúa takes the key and gives it to Sepúlveda for safekeeping.

But how many of them are there, exactly? Urzúa counts them again, and checks the list against his mental notes on how many men should be there among them. "Thirty-one, thirty-two, thirty-three . . ."

"There are thirty-three of us," he announces.

"Thirty-three?" Sepúlveda shouts. "The age of Christ! Shit!" *¡La edad de Cristo! ¡Mierda!*

Several other men repeat the phrase, including Aguilar and Lobos. *"¡La edad de Cristo!"* they yell out. Even for men who aren't especially religious, the number carries an eerie meaning, especially for those who have reached and passed the age themselves. Thirty-three, the age of a crucified prophet. The number and the name sit there among the group for an instant, a coincidence that's both trivial and frightening. Really, there should only be sixteen or seventeen of them, but thanks to all the men working overtime, or makeup days, there are many more. Twice as many, in fact. So many that no one man has met all the others. Thirty-three in all. How can that be?

Finally, Sepúlveda speaks, loudly, because in the eyes of the men around him he sees confusion and fear. *Somos treinta y tres.* "There are thirty-three of us. This has to mean something," he says. "There's something bigger for us waiting outside." He says this with the anger of the street fighter he once was, and with the conviction of the father he's become, a man who's seen a wall of stone and a half-empty cabinet of food, and who refuses to believe it marks the end of his life's journey.

One group goes back up to the rock at Level 190, and to the nearby chimneys and caverns, to listen for the approach of rescuers and then to make noises alerting people on the surface to the presence of living men down below. They'll be so busy moving rocks, lighting fires, and doing other things that they won't be able to sleep for a couple of days. But most of the thirty-three trapped men stay in or near the Refuge. A few, in fact, are afraid to leave that room, and won't for several days, because they can't forget running for their lives in the collapsing, exploding mountain outside. Sleeping behind the steel door of the

Refuge, or just next to it, they can at least pretend they are in a safe place.

"Remember those Mexican miners who were buried underground," one of the miners says. "They just put a stone over the entrance to the mine and said, 'They're dead. This is their tomb.' They didn't even bother getting out their bodies."

"No, you're wrong," another miner shouts back. "Right now, our families are all up there. They're going to make sure they come after us."

Someone says that the rescuers could carve a new ramp to come and get them. Maybe even a ramp from the brother mine nearby, the San Antonio.

"But it took ten years to make the ramp that's here now," says Yonni Barrios, who's worked at the mine that long. "It would take ten years for them to reach us that way."

Or maybe we could climb out through the Pit, another miner suggests.

No, that would be suicide, useless, like trying to climb a cliff of shifting and tumbling boulders, several men reply. You'd be sure to fall or be crushed.

One of the older miners says the only solution is to drill for them. A drill can reach them in a few days, send down food, keep them alive while the people on the surface devise a rescue plan.

So they'll reach us in a day or so, someone says, hopefully.

No, another miner answers. "Did you see a drill out there when we came into work today? No. They're going to have to bring one in from another mine. And then they're going to have to build a platform for it. It's going to take a few days, at least, just to get started."

It's past 10:00 p.m. as the men scatter about the Refuge looking for a spot to sit or lie down. There is nothing else to do there, for the moment. Several make beds from the cardboard boxes that once stored various explosives, or with the soft plastic ripped out from the ducts that pumped fresh air from the surface. On a normal day they'd be back in their bunks at the hostel in Copiapó, their bellies warmed by wine or beer or something stronger, or in their homes dozing off to sleep with wives, girlfriends, children around them. This is the hour when their bodies usually give up, surrendering to gravity and sleep after twelve hours on their feet, underground, but tonight the only rest is on the white floor of the Refuge, the sandy surface of the Ramp, catching the eyes of other men, exhausted, disoriented, with the childlike stares of the lost. Ten years for a new ramp to reach them. Days just to hear a drill. Or maybe just the silence of being forgotten, of having that stone across the Ramp be the closed door of their tomb. When there is nothing left to say, they open their eyes wide in the darkness and think how cruel and how wrong and how unfair it is to find themselves here, among these other sweaty, smelly, and frightened men.

There was comfort in the rushing rhythm of their daily working lives, in the 8:00 a.m. to 8:00 p.m. shift going underground and coming out again, and then back into this mountain where the tiniest share of the bounty of copper and gold was theirs to keep. Now there is nothing to do but sit still, listen to the intermittent thunder of falling rock, and wonder if this is all they will know. Maybe all the pleasures of sweat and simple living under the sun, the moon and the Southern Cross in the night sky belong to the past. So many memories left behind, out there in the world of the unburied: filling boxes with picked grapes, seeking out the pretty newcomer at the family gathering, joining friends for hard drinking, walking through doorways and into Copiapó bars where stationary mirror balls awaited. Collecting paychecks and coming home at 9:00 p.m. to the voices of children gathered under the streetlamps in the sloping neighborhoods of Copiapó, under lights tinged amber and emerald. The outside has slipped into the *was*, because now they live in a present, and perhaps a forever, of darkness. The past was family patios where men gathered to discuss whether La U or Colo-Colo would win the next *fútbol* championship, and other important and relaxing subjects of male-centered conversation. The past was the open windows leading to their backyards, to grills and the cracked skin of cooked sausage. It was the silhouettes of their pregnant wives and girlfriends, moving about living rooms and kitchens, the mystery of the feminine there in the bellies growing with their progeny.

Two of the miners are awaiting the birth of their children. There is Ariel Ticona, a spry twenty-nine-year-old, who already has two with the same wife. Richard Villarroel is a tall mechanic. His pregnant *polola* is called Dana, and he lives with her in Ovalle, several hours to the south of Copiapó, a place that fancies itself as a kind of Eden, a haven of palm trees and flowing waters amid dry, barren hills. Tonight his girlfriend is a pregnant Eve in that oasis while he, her Adam, is stuck in a hole paying for their recent carnal sins. He remembers her swollen belly, and the baby swimming inside, and those first few faint kicks he felt when Dana brought his hand to that hard shell of skin. Those kicks, he now realizes, might be the closest he ever gets to knowing his son. Richard's own fisherman father died when he was five, in an accident on a lake in Chilean Patagonia, leaving Richard with a lifetime of unsettled thoughts and shifting homes, and finally a teenage rebellion against his widowed mother in which he actually ended up in jail, briefly, angry at the unjust world that would deprive a boy of all but the faintest memory of having a father. It was as if his father's life had been taken by a lightning bolt, and Richard's death will feel that way too, to Richard's son, if he dies here. It's an act of chance, the absurd hand of fate at work, because Richard wasn't even supposed to be underground. He signed up to work aboveground, and he knows that his mother will be confused when she sees his name on a list of missing men, because as far as she knows, he doesn't even work in a mine. The idea that Richard will soon leave his son an identical legacy of absence to the one he knew, a lifetime of suppressed suffering, now looms over him.

That's the cruelest thing about this August 5, a day whose final minutes are playing out in the Refuge with the sounds of men moving about on their makeshift resting places: the knowledge that they will be missing to all the people who love and hate them, who depend on them and who are frustrated by them. They won't be there to protect or provide, to be fed, to be listened to, or to hear the complaints of mothers-in-law or face the angry wordlessness of their adolescent children. The miners won't be there for the party for the baby, *la guagua*, or at the cemetery to place sunflowers and marigolds on the graves of the men who raised them to be something other than a miner, someone other than a man who could die this way.

Omar Reygadas, the scoop operator who was at the bottom of the mine when it collapsed, was at the cemetery in Copiapó just a few days earlier. He was in the sunshine, walking past the roofed portal into a space cluttered with crumbling crosses and monuments, the nearby dun mountains visible in the distance. He's a widower, and during these last seven days off he had gone to see the grave of his late wife, the mother of his children, a woman he left while she was still alive. And next to her, the grave of their adult son, who died in an accident. During those same days off, he had gone to the lawns at El Pretil Park, under the pepper and eucalyptus trees, for a party and barbecue in honor of his seven-year-old grandson, Nicolás. "All my children were there, my grandchildren, my great-grandchildren." And he had gone to Vallenar, the town where he had grown up, an hour to the south of Copiapó, to see his brothers. All of this during just a few days of rest, perhaps his last. Remembering this, and being of a sentimental age, it now looks to Omar as if all these events were foreordained: as if God had given him the chance to say goodbye to everyone before leaving this Earth. It's a thought at once comforting and devastating, because it means facing the fact that this really is his end.

Omar thinks: *God, if you're going to take me now, at least let them find my body eventually.* He begins to weep. "I'm not embarrassed to say I cried, a lot, at that moment, thinking that I wouldn't see my family again, and thinking of the suffering they would go through outside." He doesn't want his fellow miners in and near the Refuge to see him broken, so he steps out of that place and begins to walk, alone, in violation of a mining code that says you should never walk alone underground. Safety doesn't matter anymore, the rules don't matter, and he walks downhill, following the light of his lamp, until he finds a front loader like the one he used to operate. He sits inside the cab, a quiet place to think, but after a few minutes he remembers the moment of the collapse. Tons and tons of rock fell on top of them, and yet "there wasn't anyone who was hurt, not even a scratch." They are thirty-three trapped men, suffering with their fears and their memories, yes—but they are *alive*. Omar realizes that the improbable fact of their survival also carries a hint of the divine. To be alive in this hole, against all odds, speaks to Omar of the existence of a higher power with some sort of plan for these still-living men. He decides to go back to the Refuge, and to wrestle with his own fears and to be a strong old man instead of a weak one. He thinks that maybe if he can transmit strength to those men and boys up there in the Refuge, that will be good for something. And if it's all part of a plan from his Creator, then maybe Omar's prayers, his thoughts, and his will can reach up to the surface, too, and make the people who love him strong, because they must be suffering, out there in the night, wondering if he and thirty-two other men could still be alive.

5

RED ALERT!

Down in the Refuge, Darío Segovia, the miner who said goodbye to his girlfriend with a long, silent embrace, says very little as the hours pass from night into morning. Darío has crevices that cut across his sandpaper cheeks, and his expressive face speaks for him: He might narrow his eyes to say *I am determined*, or allow his otherwise steady brow to retreat into a silent statement of worry. During his first twenty-four hours trapped underground, a wrinkling of his leather brow says, very clearly, *I am confused and afraid*. He's always been a man of few words, as his big sister, María, can attest. She spent much of her childhood looking after her little brother and four other siblings, and like the rest of the Segovia family she knows him by his middle name, Arturo. When Darío Arturo Segovia was a teenager he worked in a mine carrying rocks in a leather bag called a *capacho* that slipped over his torso. His relatives smiled and laughed at the eagerness with which Darío accepted his job as a beast of burden, and they made "Capacho" his nickname.

When they were little, growing up in the Atacama Desert, Darío Arturo spoke little and María spoke a lot, defending her little brother and their siblings against the vicissitudes of a life with parents who often left them to fend for themselves. Today all the Segovia siblings are deep into middle age, approaching or just past the landmark of fifty, but María, small and squat and sunburned from many days and years working in the open air, remains the one who takes charge in a crisis. So it will be today, even though she is on the other side of the desert, more than three hundred miles from Copiapó. María Segovia is at the city hall of the port city of Antofagasta, and she's there for the same reason Darío is in the mine—trying to win one of the small victories that give her a dignified existence. Her life has been this way for as long as she can remember. When she was a girl of about nine taking care of Darío, who was six, they lived in the desert town of San Félix, in a home built of river stones, wire, and wood scraps in a ravine. La Quebrada de los Corrales, the ravine was called, because there were corrals for animals nearby. Drops of water began to fall from the perpetually dry skies, drumming on the nylon tarp that was their roof, followed by rivulets that snaked along the floor in the spaces where they slept and ate. A torrent knocked down the stone walls of their home and carried away their belongings. Later, María was pregnant at fourteen, and today, at the relatively young age of fifty-two, she is a happy great-grandmother, but memories of poverty, of living in places that can be literally swept away in a moderate rainstorm, lead her very often to be a pain in the ass to anyone who gets in her way.

María Segovia has been bothering the people in Antofagasta City Hall for years for the various permits required to sell ice cream and stuffed empanadas on the street and by the beach, from baby carriages and from metal carts. If you don't have the proper permits, the Carabineros can haul you away, and María knows what it's like to be tossed into a jail cell for the crime of selling baked snacks. She's in line to renew her permit to sell candies when her cell phone chimes with a call from the wife of her brother Patricio.

"Hey, María, has Patricio called you?"

"No."

"You don't know what happened to Arturo?" her sister-in-law asks, referring to Darío by his middle name.

"No."

"I think you should call your brother."

A few moments later, Patricio breaks the news: Darío Arturo had an accident in the mine. He's trapped.

María gets off the phone and decides she needs to find out more, and fortunately there are computers in city hall for public use for her to search the Internet. She finds a news report about the accident, followed by Darío's name on the list of trapped workers, and finally, and most disturbingly, his face there on the screen. Darío has been working in mines since he was very young, and standing there staring at his face, she thinks the accident must be a very serious one to leave him without the means of an easy escape—because Darío always manages to get away and come home.

A few minutes later, María is walking out of the Antofagasta city hall, headed to the bus terminal for the long ride south. The bus takes several hours to cross the Atacama Desert. By 4:00 p.m. she's arriving in Copiapó. She heads out to the hospital, where miners' wives and children and girlfriends are all waiting for news, but of course there isn't any. María decides she has to travel to the mine and get as close to her brother as she can, but one of the hospital employees tells her: "They're not letting anyone go there. It's all closed off. You have to wait here." They say this again and again—"You have to wait here . . . You have to wait here"—but their insistence only makes her more determined to get closer to her brother. She calls her son Estebán, who lives in Copiapó.

"Of course, Mami, I'll take you there," he says. But first he takes her to his home, and she prepares by getting a good strong coat and blankets, because her son says, "Mami, it gets cold out there where the mine is." She takes a thermos of coffee and sandwiches, too, because she knows she's going to have to wait, because more than likely she'll have to face another trial of patience. That's been the lesson of María Segovia's life up to this moment: You defend your humanity with patience and determination, by making your voice heard to those who judge you a lesser being for your timeworn clothes, your callused hands, and your sunburned skin.

Duly wrapped against the cold, María arrives with her son at the San José Mine at around midnight, almost thirty-six hours after the accident. She sees families gathered around bonfires, wandering about the dusty, dry ground, or sitting on piles of gray rocks the size and shape of bread loaves, the worry thick on their faces, staring into burning flames, standing with their hands in their pockets underneath posts where cones of weak light are swallowed by the immense blackness of the Atacama night. The mine presents an instant sense of collective tragedy, of loss that's spread across the landscape like some kind of infection. "*Un panorama horrible,*" María Segovia says later. "I can still feel that sensation in my body, that grief, that feeling of being sick to my stomach, the worry that came from having my brother close to me, and having him be far away in this terribly deep place."

There are fire trucks present, and a few police officers, but no rescue taking place, not yet. She finds a major in the Carabineros, Rodrigo Berger, and he is very polite and respectful but has little information to offer. Everyone is waiting for a new rescue team to assemble and enter the mine. All have come and will come from other mines, because the San Esteban Mining Company itself has no rescue crew. Several dozen helmeted men linger before the mine opening, and all the nearby relatives can do is wait. Wait, wait, wait as each minute passes, living with a kind mourning, because the idea that the men may already be dead, or dying, is sitting there among them all, unspoken, the possibility they've learned to deny all their lives, suddenly here, so real in that mine mouth which even the hardened miner-rescuers seem reluctant to enter.

Darío Segovia is down below in the Refuge, saying very little. On the surface, on this cold night, his big sister María feels that he is alive. Today, when they are both middle-aged, she will defend him, just as she did when they were children in San Félix, in a home that flooded when it rained hard. Soon she will begin speaking, telling everyone what she believes: Darío Arturo "Capacho" Segovia and the other thirty-two men are still alive, and they need their families to fight for them.

José Vega is a fit, wiry man of seventy with smoky brown skin and curling sideburns who began to work in mining when he was a teenager. All four of his sons have worked in mines, though by the time the August 5 collapse hits, three of those boys have mercifully quit that dangerous work. Only one, Alex, is still working underground.

"We're going to rescue him," José tells his adult son Jonathan. "Let's get our equipment and get ready to go inside."

José gathers the equipment he has left from his working days: a compass, a GPS receiver, a device to measure depth, an emergency oxygen device. Two of his sons will join him in the search for their brother, meaning that three Vega men who left mining are going back into a mine to save the one member of the family who was too stubborn and hard up for money to quit. When the Vegas arrive at the San José they find police officers, firemen, and groups of rescuers from many other mines. The rescuers are going to enter in groups of six. But suddenly someone announces: "We're not going to let any relatives of these guys go in there." When it's his turn to sign in, José Vega gives a false name. They're told they're going to have to wait for several hours before entering, and since Jonathan looks very tired, his father says he should go home and take a nap: "We'll send for you when it's our turn to go in." But when José gets word at 4:00 p.m. on Friday afternoon that he'll soon be going in, he doesn't call Jonathan's house to have him woken up. "I already had one son in the mine," José will say. "And now three more Vegas were going in. That's too many Vegas in that mine."

Other rescue teams are coming out of the mine looking worried. José talks to a group of men who've come from the Michilla mine, three hundred miles away to the north, near Antofagasta. Tears well in their eyes as they describe what it's like inside. The mountain is still alive, they say, slabs are breaking off from the walls down there, and there are huge cracks in the floor and ceiling of the Ramp. The head of the Michilla rescue team declares, "No one else should go in!" But the situation outside the mine is chaotic, it feels as if there's no one in charge of the overall rescue, and no one tries to stop the next group of men, including Alex Vega's father, from entering.

Not long after passing through that gaping mouth that leads from afternoon sunshine to darkness, José Vega begins to realize exactly what he's gotten himself into. "To be honest, it was very, very frightening." They approach the blocked section of the Ramp that Carlos Pinilla and Pablo Ramirez reached a day earlier—the rumbling mountain has piled up stones on the roadway before it. José speaks with another member of the rescue team, a relative of the miner Jorge Galleguillos, and they agree to look for the nearest chimney. When they find one, a young, skinny miner agrees to climb down inside. After being lowered down and then pulled back up, the miner reports that he could see an open passageway at the bottom. It's leading to another air duct and perhaps to a gallery where they can reach the men. But José says they don't have permission to go any farther. They climb back up to report their findings to the team of professional rescuers preparing to enter the mine.

•

The minister of mining, Laurence Golborne, arrives at 2:00 a.m. on Saturday. He, too, sees the men and women sitting by the bonfires and their expressions of confusion and sorrow. He's still dressed in the business suit he was wearing the day before while on a state visit to Ecuador. Several family members corral him and repeat the rumors they've heard: "They're dead already, aren't they . . ." When you're a government official in Chile, as elsewhere in Latin America, people believe it's your nature to misinform or manipulate, that you somehow surrender your humanity and morals when you take office. The poorer someone is, the more likely they are to feel this way about their government. "Tell us the truth, Señor Ministro!" The truth is that the available information is spotty and inaccurate. According to one count given to Golborne as he arrives, the number of trapped miners is thirty-seven. Or maybe thirty-four. He is told (incorrectly) that the ranks of the miners include several illegal immigrants from Peru, or maybe from Bolivia.

Not much in Golborne's professional career up to this moment has prepared him for this. He's a well-off executive and business owner with a degree in civil engineering and a minor in chemistry who's never before held a government position. Upon taking office he was more familiar with the challenges in serving South American haute cuisine (he owns a restaurant in a tony district of Santiago) than in the complexities of the mining industry (his only experience in mining was as an administrator, a number cruncher, twenty years earlier). His journey to Copiapó has been a long and roundabout one after leaving the state visit with President Sebastián Piñera to Ecuador, flying coach on a commercial flight back to Santiago, and then in a Chilean air force jet to Copiapó. Golborne is part of the circle of businessmen, politicos, and thinkers now taking the reins of government after two decades of center-left rule, and at this moment his business attire isn't the only thing that makes him stand out. He's a top-ranking official of a conservative national government in a region that voted overwhelmingly for the left, and his presence at the mine is unusual because the federal goverment never takes a role in mining accidents—by tradition, it's usually the mining companies that lead and organize their own rescue efforts.

"The attitude among people there was, 'Well, you're here.' They weren't openly hostile, but they weren't espeicially welcoming either," Golborne will say later. He watches as the rescuers gather near the entrance to the mine; they are from Punta del Cobre, Escondida, and other mines. Just after midnight, they pass underneath the jagged gray teeth at the entrance and disappear inside.

The rescue team is composed of Carabinero police officers and a local mining rescue team lead by Pedro Rivero, a lifelong miner with a ponytail dyed blond and many years' experience in mining rescues. Off duty, at home, Rivero is a transvestite, a fact widely known in the local and eminently macho mining community, where he is nevertheless respected for the undeniable manliness of his courage as a rescuer. His team includes one employee of the San José Mine, Pablo Ramirez, the night-shift supervisor and Florencio Avalos's friend. They drive the 4.5 kilometers into the mine until they reach the flat granite wall that's blocking the Ramp, then assemble their equipment for a descent into the deeper reaches of the mine, hoping to reach the Refuge, some 285 meters farther down. They bring ropes, pulleys, tackle blocks, and slabs of wood for building a platform over one of the chimney openings, each of which is 2 meters (6.5 feet) wide, and as much as 30 meters (100 feet) deep.

Rivero, Ramirez, and five other rescuers descend through several chimneys from the spot where the Ramp is blocked at Level 320, down to Level 295. To Rivero, who's never been in this mine before, the scene is increasingly "Dantesque" the deeper he goes, the more the temperature rises and debris fills the Ramp and cracks begin to appear in the structure of the stone. At each level they find the same perfect gray mass blocking the Ramp, and stop to yell downward, in the general direction of the trapped men, calling out with the usual insults with which mining men address one another: "*¡Viejos culiados! ¿Dónde están?*" With each level down, the sense of danger deepens. They are like a Himalayan expedition working in reverse, their goal to "assault" the center of a mountain instead of its peak, with the air getting thicker and hotter instead of colder and thinner, and at Level 295 they gather to make a kind of base camp. They assemble more ropes for the descent into the next chimney, which leads down to Level 268.

Ramirez goes first, alone, and after a few meters he begins to notice something deeply disturbing: There are cracks in the walls of the chimney. The stone walls of the cavity are coming apart, being slowly squeezed by the weight of the skyscraper-size "mega-block" of stone that's fallen from above. Once again he yells down below for the trapped men and listens for a reply that never comes. When he gets near the bottom of the chimney he can't see the Ramp at Level 268; instead, there is a pile of debris, a hanging wall of rock that looks like it could collapse at any moment and seal the opening completely. "Anyone who passed through there," Ramirez later says, "would be in danger of being stuck if all those rocks came tumbling down." This isn't going to work, we can't go through here, he thinks, it's going to crumble very soon. Ramirez makes this assessment in his thoughts, and feels suddenly defeated. He thinks about Mario Gómez, the truck driver, and the men in the mechanical crews, and like Pinilla he's certain they got caught in this collapse, because it was the lunch hour and they had to be driving up the Ramp near this spot. But the rest of the men are probably alive down there, Ramirez thinks, because in the San José the main group of fortifiers and jumbo operators never took the lunch hour when they were supposed to, they were always late or early. "That's the way mining is. You always find a way to cheat fate," he says. "That's what's beautiful about being a miner: Supernatural things always happen."

Now the weird, mysterious thing is the mountain rumbling all around him, from rockfalls that may be distant or may be close. He yells up to Rivero at the top of the chimney.

"¡Sácame, huevón, esto se fue a la cresta!" Pull me up, you bastard. This is going to hell!

Rivero and the other rescuers at the top of the chimney also hear the sound of thunder rumbling from the stone around them. A huge slab of rock comes falling down from a great height, and it severs the cable linking their communication equipment to the top. It's like hearing a bomb go off nearby. Rivero begins yelling: "Red alert! Red alert!"

Rivero and the other rescuers pull up Ramirez, but the wall of the chimney is starting to crumble, and one or several slabs catch the loose ropes the rescuers have attached to their bodies to follow Ramirez down into the chimney. For a moment, it's as if the mountain were trying to suck them down, because falling stone is pulling at the ropes, trying to drag them into the chimney, pulling so hard that Rivero loses his footing. He and his fellow rescuers fight to keep from being sucked into the chimney even as they struggle to pull up Ramirez. When Ramirez's head emerges at the top, they grab his arms and lift him while another rescuer pulls a knife from his boot and cuts all the ropes. Only at that instant are they safe.

The men begin their journey back to the top, with Rivero noting how disconsolate Ramirez looks, worried for his friends and coworkers below. Rivero thinks, *This is over now.* At 3:00 p.m., Rivero and Ramirez reach the top and walk out into the Atacama sunshine. Word of the collapse inside the mine has already spread to the rescuers gathered outside, including José Vega, Alex's father. "There's millions of tons of rock and it's impossible that they're alive," one of the rescuers gathered nearby says.

"It was total silence after he said that," José will recall.

Golborne is there waiting for the rescue team, too, and the minister is at once struck by the their defeated expressions, and by their reddened eyes and the paint of gray and black grit covering their faces, as if they'd gone inside the mountain and been fighting some beast that lived there. *"Esto se acabó,"* Rivero says, with an unavoidable sense of finality. There is nothing more Rivero or any other "conventional" mining rescue team can do, and now it falls to Golborne to tell this to the waiting families.

As the highest-ranking official present, Golborne has taken on the strange role of spokesman in the hours since arriving, even though he doesn't have any legal authority at the mine (techinically, he's trespassing on private property). But the mine owners aren't talking to the media or the families, and the reporters drift to the handsome, educated official. The families have been waiting two hours to hear from him, and now he walks away from the mine entrance to an ambulance parked near the main gate. The minister stands on a chair placed next to the vehicle and uses its loudspeaker to address the crowd, television cameras transmitting his statement live to all of Chile.

"The news is not auspicious," he begins. The rescuers were trying to reach Level 238, he tells them, but there was another rockfall. The mountain has collapsed again, and the rescuers have had to flee. Several family members gasp at the news. "We're trying to find other techniques, other mechanisms to reach them," Golborne says, but as he utters these words he looks down and sees Carolina Lobos, the adult daughter of

Franklin Lobos, the former soccer star. She is looking distraught, help-less, and her eyes are filled with tears, seemingly more tears than a pair of eyes could possibly hold (that's how Golborne will remember the mo-ment). At the sight of her tears he stops speaking and looks down and away from the crowd. He feels "this knot in my throat" but manages to resume speaking for a few seconds. "We will try to keep you informed . . ." he announces, but then he has to stop again, because it's simply too much, to be the official to tell so many good people, so many daughters and wives, that the men they love are trapped and that he doesn't know how or if he'll be able to save them. "We have to take care of the rescu-ers' lives," he says, and a few words more, and finally he feels the emo-tion welling up again, and he turns away from the crowd completely, and gives up and puts the microphone down.

"As you can see, the minister broke down, visibly shaken," the an-nouncer on Chile's TVN news network says in a grim tone. A sense of shock and mourning spreads across Chile, because ministers don't cry like that on live television, and seeing one do so makes the moment all the more real and extraordinary for the millions watching. Something very tragic, very human must be happening for a powerful man to cry like that.

Golborne steps down into a crush of reporters. They're asking him questions, and he gives them only vague responses about studying "op-tions" and other "techniques," and as he speaks the microphones pick up the sound of a woman, or many, weeping very loudly. The "short-term" solution of going through the chimney won't work, Golborne says, al-most mumbling now.

In response a man nearby shouts: "Señor Ministro. We can see it in your face. Tell us the truth! Can you get them out? We're all in the dark! We've been waiting here fifty hours. Fifty hours!"

Golborne says that "hopes have to be realistic . . . We can't transmit an optimism that doesn't exist."

Hearing this, a relative and miner from Cerro Negro shouts: "David took on Goliath! And he used the weapons that he had!"

"You can't break down like that!" a woman tells him. "You're a minister. You're the authority here. You have to show you're in charge!"

Golborne is up to this point in his career a Stanford and Northwest-ern alumnus with a stellar corporate résumé, but he's never been someone who's had to think of himself as a man of the people. He's never had to make the concerns of the poor his own, and he's never known what it's like to be the public servant of people who want him to be strong and who are deeply suspicious of him at the same time.

María Segovia remembers what happens in the minutes and hours that follow this way: She and the other family members watch, in the fading afternoon light, as rescue-team members walk away with their helmets underneath their arms and fire trucks drive away from the mine entrance. She feels abandoned, and so do most of the family members around her. They're weeping, she's weeping, but finally María stops and thinks: *We're not going to get anywhere with tears.* Eventually she finds a young reporter for CNN Chile, and she tells him, "No, this can't be this way." She describes seeing the rescuers leave, and speaks of her outrage and hurt. "These men are not dogs that are buried here," she says. "They're human beings. And for that reason we need help. Because they have to be rescued." Call the president, she says, call other countries, bring help to get these men out of here.

In the days to come María Segovia and the other women who've come to the San José Mine will often feel the need to speak and protest. The police tell them to move away from their spot near the mine entrance and force them back down the hill, outside and past the front gate: "We retreated like dogs with our tails between our legs," one woman remembers. When María and the other family members don't get information, they walk back up to the front gate and start banging pots and pans. It's the only way. They'll block the road and the entrance to the mine if they have to, and they do so more than once. Each time they stage these impromptu protests, they face off with several police officers lined up there to keep them off the mine property. "We want information!" the women yell. "Come down here and tell us what's going on!" A short time later a government official comes down and talks to them. Carmen Berríos, Luis Urzúa's wife, sees Alejandro Bohn, one of the mine owners, walk past, and confronts him: "You don't have a secretary to call us about the accident! I had to find out on a miserable bus!" He mutters something back and walks away.

María Segovia quickly realizes that the most important thing she can do to help her brother is to simply stay there, at the mine. "We're going to have to live here," she tells the other women, the mothers and girlfriends and daughters. "We're going to have to stay here until the bitter end." *Hasta las últimas consecuencias.* María sets up her tent closest to the gate, and as the camp of family members around her grows it soon comes to have a name: Campo Esperanza. This Camp Hope becomes a draw to charities and well-meaning people from across Chile and beyond. People begin to drive into the moonscape desert with donations for the miners' families. More firewood. Tarps. Food. The regional government in Copiapó brings tents. A small chapel goes up. Eventually, the government will even provide a school for the children who are living there with their parents.

María Segovia is not the only one taking charge, she's not the only one who's decided to live at the San José Mine, but she's the one closest to the gate, the one who most often answers questions. A truck arrives with more firewood and the driver asks, "Where do we put this?" "Over there," María says. "We'll make sure everyone gets some." The government officials who go down to talk to the families very often encounter María first, and they are moved by her determination, by her love for her brother, her unflagging defense of the dignity and humanity of the men trapped down there. Soon this woman who sells ice cream and candies and empanadas from a cart at the beach will be on a first-name basis with some of the most powerful men in Chile. "Don't forget my brother, don't let those men die down there," she tells them. As the community of relatives and rescue workers grows around her, María speaks so much that this new community will finally give her a nickname: "La Alcaldesa." María Segovia will become the "mayor" of Camp Esperanza.

But much of that will happen in the days to come. In the short term, a day or so after the families of thirty-three buried men first began to bang pots and pans, after they first made the minister of mining weep, something very important happens. María Segovia and the other family members stand before their camp at the entrance to the San José Mine and see a strange vehicle come rumbling toward them.

6

"WE HAVE SINNED"

Six hundred meters below the fires of Camp Esperanza, in the damaged mine down by the Refuge, Víctor Zamora and a few of his companions walk out onto the Ramp and begin to look for Víctor's missing teeth. The beams of their headlamps pan over the gravelly ground. They are surrounded by the complete and, yes, terrifying darkness of the mine, and they're starting to think about the batteries in the lamps going out: They begin each shift fully charged, but already a few of the miners can see the beams dimming. Turning on their precious lights, they look, briefly, thinking the pearl-colored beads of Víctor's teeth will catch the light if they're here. But they find nothing.

A bit higher up the mine, Luis Urzúa, Florencio Avalos, and the mechanics Juan Carlos Aguilar, Raúl Bustos, and others travel back through the black tunnels uphill in Urzúa's truck, driving up to the rock wall that's blocking the way out. They listen for a sign that rescuers are near, or headed toward them. In the darkness every sound and sensation is amplified: The faint circulation of the air through the mine starts to feel more like a breeze, the breathing of the man next to them sounds like the rustling of an unseen animal or a passing vehicle. But they hear no sound that is clearly that of a rescuer.

Urzúa has surrendered the authority of his white helmet, but he's still making suggestions, responding to the ideas of the other men, who collectively are trying to take charge and keep busy. Juan Carlos Aguilar, Mario Sepúlveda, Alex Vega, Raúl Bustos, Carlos Barrios, and Florencio Avalos: They all want to do something besides wait. Ideas are suggested, the men debate, and Urzúa and Avalos share what they know about the mine, and together supervisors and underlings reach a decision that makes the most sense. It's not the way things are done during a regular workday but in the absence of a single man taking charge it's what seems best. At the base of the chimney where Mario and Raúl climbed to seek a way out, they set fire to a small tire from a wheelbarrow, and also to an oil-soaked air filter from one of the machines, hoping the smoke will drift up and reach the surface, sending a sign that there are living men below. But the smoke simply gathers around them in the Ramp in a useless cloud. There are hours when, for reasons no one can explain, the very faint breeze at the workshop by the Pit shifts, and so they try again when the air is circulating upward into the cavern, though the movement of the smoke is slow. They watch a wispy cloud disappear into the cavern beyond the short reach of their beams and are not hopeful it will reach the top. Next they stuff a detonating cord inside one of the rubber tubes that run inside a chimney and set it alight. These tubes carry phone lines and electricity and compressed air, and in theory it should be possible for the pungent smoke produced by the burning cord to travel up that tube to the surface; any miner who detects that scent will know what it is, and know it's a message from men who are alive below. Of course, the tube was more than likely cut somewhere by the same collapse of stone that's trapped them, but the men decide to try anyway, and the smoke does seem to stay inside the tube and drift upward, though it's impossible to know how high it reaches. They go to one of the galleries filled with rocks, because some of the miners think that if they clear that space there might be a way to climb to another opening that leads upward. Mario uses the big scoop of a front loader to lift out the rocks, but it's a Sisyphean task, because more rocks fall from the top of the pile to take the place of those he clears out. "I practically destroyed that loader," Mario will remember. They consider building a ladder to climb the chimney, using rubber hoses and pieces of rebar, but realize it likely won't hold a man's weight and that the one saw they have won't be able to cut more than a few pieces of rebar anyway. Later they bring up one of the vehicles to the stone guillotine blocking the Ramp and honk its horn, and they use one of the jumbo machines and pound its arm against the stone wall. Then they stop and turn off their lamps and listen to the quiet, again, to see if they can hear a reply coming from the blackness around them: a honk, a clank, a banging, perhaps. But there is nothing, just the stone blocking their way, their ears turned toward it.

While a dozen or so men are doing this work, the majority of the miners remain below in the Refuge. "They weren't bothering anyone, and that was good," Urzúa later says. "As far as I could tell, they weren't fighting." But the men are already being divided into "doers" and "waiters." The doers will do anything to fight the idea that they're trapped and dead, and they believe the waiters are afraid, and that they won't leave the Refuge because they're paralyzed by the memory of running for their lives with the mountain falling around them.

There isn't much that makes the small space of the Refuge safer than the rest of the crumbling mountain around it: There's the steel door, and strands of chain-link fencing covering the stone walls inside, a kind of steel net that's supposed to keep the rocks from crushing the men inside if the rest of the mountain crumbles. The Refuge is inside the same mountain as the Ramp and the rest of the broken mine, but it does have a blue-and-white sign outside with the words REFUGIO DE EMERGENCIA. Inside there's the locker with their remaining food supplies, a first-aid kit on the wall, and a single picture of a naked woman ripped from a magazine. For the first few hours after the disaster the Refuge stays as neat as Franklin Lobos left it for his boss, Carlos Pinilla, to inspect, and *los niños* respect the blue signs telling people to throw their trash in the bins. A small digital thermometer displays the temperature, 29.6° Celsius (85.3° Fahrenheit).

At noon on the second day, all thirty-three men are present as Mario Sepúlveda divides and distributes their daily "meal." He lines up thirty-three plastic cups in rows and spoons one teaspoon of canned fish into each cup, then pours in some water, making a bit of broth. He passes out two cookies to each man. "Enjoy your meal," he says. "This is delicious stuff. Make it last." The men linger about, a few form a kind of makeshift line. That single meal, at noon, likely contains fewer than 300 calories and is meant to hold them all until the next noon.

Several times during those first few days trapped underground, the mountain rumbles as if it's going to explode again and more people decide to sleep in the Refuge, or just outside. "I tried to stay outside, but outside I always slept with one eye open, and when the mountain made noises I'd go running back inside," Lobos says. Soon there are twenty prone men sweating in a space the size of a large American living room. A few of the men take the plastic stretchers that were stacked neatly inside and use them as beds, others toss cardboard onto the floor, and find boxes to use as nightstands. It was always a struggle for Lobos to keep that place clean, because the men would come in there to rest, to lie down on the floor with soot-covered bodies; now they sweat all day long and their moist bodies are painting the room's white tile floors with the gray grit and black soot that's clinging to them. The Refuge starts to fill with the smell of their sweating, unbathed, manly bodies—"we didn't have water we could spare to clean our private parts," as one miner puts it—and in a space where there is no longer any ventilation, the fetid scent begins to gather and cook, transforming the air into a stew of body odor. "I've smelled corpses before, and after a while, it smelled worse than that," one miner says later.

Men who sweat crave water. The few liters of bottled water in the emergency supplies in the Refuge were devoured in a day, so now they drink some of the thousands of gallons of water stored inside the mine to cool off the drilling machines. This water came down from the surface in hoses that led to a series of underground tanks, all the way to the deepest reaches of the mine. On the second day after the accident a few of the men open one of the spigots to wash themselves off but the water is too precious to use that way now—to conserve their limited supply, Juan Carlos Aguilar tells Juan Illanes to cut the hose from a holding tank on a higher level and seal it off so that no one can take a bath farther below.

Now the men fill plastic barrels at this tank. Mario Sepúlveda organizes them in teams of three, to drive one of their vehicles every two days to the tank and fill up a 60-liter (16-gallon) barrel with water. They put this water in their plastic bottles and look at the dirty liquid inside, and think about how it's keeping them alive. Before the collapse, the men would wash off their dirty gloves in this water. Sepúlveda used to jump into the water tanks to take a bath, in his irrepressible, impulsive way, and a few men come to the disgusting and comical realization that they're all drinking Mario's bathwater. When they shine their weakening lamps on the water in their bottles they can see a thin, black-orange film and small drops of motor oil. One of the miners thinks the water tastes the way a pond smells when ducks defecate all over it. As disgusting as it is, however, a few sips can make their hunger go away.

The hunger hits them worst these first few days. It creeps up on them. Suddenly they can't even go to the bathroom to defecate, even though their bodies seem to be telling them to try, because for many the emptiness in their stomachs is like a fist pushing the emptiness downward. Franklin Lobos used to be a professional athlete, attuned to the state of his own body, and perhaps for this reason as he sits in the Refuge he begins to assess the state of the men's health. Mario Gómez, the truck driver with two fingers missing, is clearly the worst off. He's got that silicosis cough that won't stop. You hear that cough and you hear history, as if the cough were something passed down from his miner grandfathers. *Is this* viejo *going to make it?* Lobos wonders. The "old man" can't possibly make it. José Ojeda is diabetic: Is two cookies a day going to be enough to keep him from going into shock? After a day or two, Víctor Segovia breaks into a rash all over his body. Is it the heat, or nerves, or both? Jimmy Sánchez, the youngest of the miners, is acting like an old man: He simply won't get up, he's defeated by an emotional and physical lethargy that quickly spreads to other men.

The thing to keep the men from feeling hopeless is to talk, make jokes, tell stories, imagine what the rescuers are doing. Yonni Barrios fills the silence by explaining the structure of the mine to some of the younger, more inexperienced guys, including Mamani and Sánchez, drawing a map on a piece of paper for them. "Look, here we are at Level 90," he tells them, "and we can walk up to 190, and from there there's a passageway to 230, and then to 300, and then to 400."

"So we're free!" Víctor Zamora, the former "gypsy" from Arica, calls out happily in the darkness. "We can just climb out of here!" His wide, childlike face now brightens with a crazy comedian's grin that's framed with a head of matted, wavy hair—he's egging Yonni on, though Yonni doesn't realize it.

Víctor, who led the raid on the emergency food supply the first night, has become calmer and more composed than just about anyone around him. "We're going to get out of here," he's been telling the men. "Don't worry, they're coming for us." During an ordinary working day, the men of the A shift would be teasing one another mercilessly, and teasing Yonni is Víctor's way of trying to keep the men around him loose. "We're saved!" he calls out with a big, toothy smile. "We can just climb up to 400 and then walk out of here!"

"Well, no," Yonni says. "Because here at 400 the rock is cut like a pane of glass. There's nothing to grab on to. It's impossible to climb any farther."

To this Víctor gives a look of mock surprise: "What kind of a *huevón* are you?" he asks. "We're going to climb up to 400 just to die there?" And then he bursts out laughing, and the laughter spreads to Mario Sepúlveda, who repeats "*¿Sois huevón?*" and he laughs, too, and suddenly everyone is laughing at Yonni.

Yonni Barrios, the man with two tough women in his life, doesn't care that his brother miners are mocking him. He likes seeing them laugh, because at night, when his friends are asleep or trying to sleep, they look sad and vulnerable. Yonni sees the hand of one of his colleagues flutter, and watches as a tremor runs through another's torso. He knows a bit about the lives of those trembling men, and it's obvious to him that they're going through alcohol withdrawal. They've been able to satisfy their nicotine urges by scrounging cigarette butts from the trash, drying the tobacco, and rolling it up in paper, but there's no loose liquor lying around for them to sip at and calm their nerves. Such a painful thing, to see strong men reduced to a defenseless state by the absence of their fermented and distilled daily medicines. Alcohol withdrawal symptoms usually start within ten hours of the last drink, and get worse in the first forty-eight to seventy-two hours. The other symptoms include irritability and depression, but of course, there's already a lot of that going around, with most of the irritability focused on Luis Urzúa, who conveniently isn't around to hear his underlings complain about him. "He's not worth anything," they say. "Thanks to him we're stuck down here!"

Several of the men have extra cookies hidden from their raid on the supplies the first day. They sometimes sneak away to eat them, a secret they keep from all but a few men. On that first day trapped, after Mario Sepúlveda had led the inventory of the remaining food, Ariel Ticona grabbed the cartons of spoiled milk that were tossed in the trash. It was curdled but he drank it all and it never made him sick and he can even joke with the other guys: "I'm going to last longer than all you guys thanks to that milk."

In the Refuge there's lots of time to joke around, but also to slip into a private reflection. "There is a great sense of powerlessness"—*la impotencia es muy grande*—Víctor Segovia writes in the diary he's started to keep. "We don't know if they're trying to rescue us, or what's going on outside, because here inside we don't hear any machines working or anything." Víctor is the operator of a jumbo lifter and a member of an old Copiapó family, who during his forty-eight years of life has never traveled outside the Atacama region—the town of Caldera, forty-five miles away, is the end of his known world. He got kicked out of school in the fifth grade for fighting but can write fairly well, and he begins his diary with a note to his five daughters that suggests he thinks of these written words as a message that might reach them after the stone walls around him become his tomb. Before August 5, when the San José was still a working mine, he brought a pen and graph paper in to copy down information from the gauges on the lifter, and also the duplicate forms he has to fill out when he operates it. Now he uses these materials to leave a record of what is happening to him and the men of the A shift. He begins by writing an account of the day before the collapse, when he spent an entire afternoon and evening on a beer-drinking binge with his cousin Pablo "the Cat" Rojas (who is also now trapped in the mine), the two of them mourning the death of Pablo's father and reminiscing about the games they played as kids in the Copiapó River when a bit of water still ran through it, with Víctor getting quite drunk and then stopping to eat four hot dogs on the stumbling journey home. He expected to be too hungover to make it to the mine the following morning, and had even decided to keep his bedside alarm off, but for some reason he had woken up at exactly the time he needed to, and he'd made it to work feeling no hangover at all.

In his new diary he recounts how the chief of mine operations, Carlos Pinilla, drove past a group of workers in a truck just after the mine gave off a loud explosion at 11:30 a.m., ignoring their final, pleading questions about the mine's safety. He describes the horror of the collapse itself, how the walls of the Ramp seemed to squeeze in on him. Víctor signs his name at the end of this first diary entry and tries to sleep, the stone walls and the ceiling around them trembling with the sound of distant thunderclaps. Each explosion carries the possibility that it will be the prelude to another collapse, and perhaps even a final one that will swallow up the Refuge and the steel door and net that protect the men inside.

On the morning of the third day in the mine, Víctor begins his diary entry at 3:30 a.m. by listing the names of his daughters. "Girls, sadly destiny only allowed me to be with you until the fourth of August . . . I am weak, and very hungry. I'm suffocating . . . it feels like I'm going to go crazy."

When the mine is quiet, some of the men put their ears to the stone walls, because after all the talk of drillers coming for them, listening for the sound of their rescuers becomes an obsession. "Do you hear it?" one miner might say. "I think I can hear something! Do you hear it?" Víctor Zamora says that yes, he can hear it. "I was lying," he says later. "I couldn't hear anything." But he feels responsible for keeping up the men's spirits, so he repeats: "It's really faint, but yes, I think I hear it. They're coming for us."

Yonni Barrios places his ear to the stone. "It was like listening to the inside of a seashell," he will say later. You hear nothing and you hear everything, you can imagine an ocean roiling inside that shell, and then you take away your ear and realize it's all an illusion.

The camps into which the men are divided are becoming more defined. One of the mechanics sleeping by Level 105 calls the naysayers in the Refuge "the Clan." Mario Sepúlveda is one of the few men to move back and forth between both groups. He keeps frenetically busy, talking in that upbeat, squeaky voice of his. In the Refuge, his foulmouthed soliloquies are lifting up the spirits of Carlos Mamani, Jimmy Sánchez, Edison Peña, and many others. But he has vertiginous mood swings: After being funny and inspiring to his fellow miners one moment, he'll be suddenly angry and itching for a fight the next; or sullen and quiet, lost in thought. Sitting outside the Refuge, Yonni Barrios sees Sepúlveda slip into a kind of manic, angry hopelessness. He is pacing. "He was always a little anxious. And I was watching him walk up and down the Ramp, when all of a sudden he stopped. He yelled, very loudly. 'I want to pray!'" The prone and sitting men around him are startled. "*¡Yo quiero orar!*" A few look at him as one might regard a possessed street-corner seer.

"I'm angry," Sepúlveda shouts. "I feel powerless." *Impotente.* By now the men are soaking wet and more have begun to shed their shirts, but Sepúlveda, the man with the heart of a dog, looks sweatier, grimier, and more desperate than all the others. He's tried to climb up one of the chimneys, after all, and has been working to move rocks and send messages. One miner will describe him as looking at this moment "like a commando," as if covered in face and body paint for battle in the jungle. Mario falls to his knees. "Those who want to pray, come and join me," he says. Yonni looks at him and thinks: *We aren't going to get out. Perri knows this. And he wants to get good with God. He thinks we have to talk to God and ask him to forgive us.*

Sepúlveda will remember later: "I was angry at the mine owners, because they were responsible for our safety. I was angry because it wasn't fair. I'd had such a hard life already and now this was happening to me." He is going to die, slowly suffocated and starved to death, two thousand feet underground, in a dry corner of Chile far from home, forever absent from the lives of the people who need him most.

The truth is, he'd been thinking about praying a few hours earlier, during a private talk with José Henríquez, a tall, balding man from the south who is a devout Evangelical. Mario is a Jehovah's Witness, which makes them both part of a non-Catholic minority inside the mine. They've talked before the collapse about religion, because Mario once thought he felt a ghost pass through his body at the site where the engineer Manuel Villagrán was killed. After the accident on August 5, when it was clear they were trapped, José whispered into Mario's ear: "God is the only way out of this." Now Mario has issued his angry call to prayer, and with all the surprised and amused miners around and inside the Refuge still staring at him, he turns to José and says: "Don José, we know you are a Christian man, and we need you to lead us in prayer. Will you?"

From this moment forward Henríquez will be known as "the Pastor" to his fellow miners because as soon as he opens his mouth and begins to talk it's clear that he knows how to speak of God and to God. Henríquez is fifty-four years old, and he's been in mining since the 1970s. He's survived five mining accidents, including two in southern Chile that wiped out most of the men in his shift. In one of those accidents, walls of seemingly immovable, ancient rock exploded spontaneously; and in another the silent killer carbon monoxide knocked him out and nearly ended his life. This proof of his vulnerability and mortality before fate and geology has helped send him deeper into his Christian faith, and he's a loyal member of his church in Talca in southern Chile.

"We have a certain way of praying," Henríquez says. "And if you want to pray the way we do, fine. If not, you can have someone else do it."

"Don José, let's do it the way you know how," Sepúlveda says.

Henríquez drops to his knees and tells the men they should also do so, because when you pray you have to humble yourself before your Creator.

"We aren't the best men, but Lord, have pity on us," Henríquez begins. It's a simple statement, but it strikes several of the men hard. "No somos los mejores hombres." We aren't the best men. Víctor Segovia knows he drinks too much. Víctor Zamora is too quick to anger. Pedro Cortez thinks about the poor father he's been to his young daughter: He left the girl's mother, and he hasn't even done the basic fatherly thing of visiting his little girl, even though he knows his absence is inflicting a lasting hurt on her.

"Jesus Christ, our Lord, let us enter the sacred throne of your grace," Henríquez continues. "Consider this moment of difficulty of ours. We are sinners and we need you." Just about everyone who was at the entrance to the Refuge or inside is on his knees, and they are looking pious and small before God, and small before Henríquez, who is a tall man, by the standards of both Chileans and miners. He is a man of God, and suddenly here, in this tomb, the religious severity that many of them found annoying during the everyday encounters of the A shift is exactly what they need.

"We want you to make us stronger and help us in this hour of need," Henríquez says. "There's nothing we can humanly do without your help. We need you to take charge of this situation. Please, Lord. Take charge of this."

The men kneel and pray, silently. In his unspoken thoughts, Sepúlveda recites a rambling and desperate version of the Our Father, "because that's how I learned to pray when I was a kid."

"Our Father who art in heaven . . . Lord Jesus Christ, you who are the son of our Father Creator, I thank you for all the blessings, for life, for health . . . I ask today that you protect our families, because they don't know what's happening here. And give us much strength [*fuerza*] and fortitude [*fortaleza*] to keep going, because we have to get out of here." He thinks about the cookies that are their chief remaining source of sustenance and says: "I don't know how, but find a way to feed us." Around him, Sepúlveda sees the sweating and unshaven men of the A shift, men of different faiths, joined together in poses of penitence and desperation, some with their eyes closed, others with their eyes open, praying, whispering, crossing themselves. He sees men who are still in their overalls, and men who have shed them; men who are crying, and men who look perplexed, as if they can't quite believe they could find themselves on their knees in this cavern, begging God to rescue them.

The Pastor speaks again to say the men are being tested, because they lived their lives in sin. That's why they have to get on their knees, they have to literally throw themselves on the floor (*tirarnos al piso*) and humble themselves before God. We have to recognize that we're nothing, the Pastor says. In the surface world, when they returned from the mine and showered and entered their homes, they were princes, kings, spoiled sons, well-fed fathers, Romeos. They believed their private worlds of home and family spun thanks to their labor, and that as workingmen and breadwinners they had every right to expect their world to revolve around their needs. Now the heart of the mountain has collapsed on top of them, and they are trapped by a block of stone, an object whose newness and perfection suggest, to some, a divine judgment. We have sinned, the Pastor says, and so the men speak to expiate their various sins. Our Father, who art in heaven, forgive me for the violence of my voice before my wife and children, says one. Forgive me for abusing the temple of my body with drugs, says another. When you're a boy in Chile, that's how you're taught to pray, to speak to God in the first person. The men ask to be forgiven for the moments when they betrayed the women who loved them, for their jealousies and their uncontrolled desires. They ask God to guide their rescuers to the tiny room and the passageway where they are waiting, ready to accept salvation and to begin new lives as better men.

The prayer becomes a daily ritual. They gather each day just before they eat, at around noon, for a brief sermon from Henríquez, and then later from others, including Osman Araya, a man who converted to Evangelical Christianity after a tumultuous young adulthood. The prayers and the meals are the one time each day all thirty-three of them unite. Soon, each prayer meeting will include a self-criticism session at which the men apologize to one another for their transgressions, big and small. I'm sorry I raised my voice. I'm sorry I didn't help get the water. With each passing day there are fewer headlamps illuminating their prayer and apology sessions, and those still working have a light that's a little dimmer. This is frightening, to have each new prayer take you a little deeper into what may be a final and unending darkness. A little later, Juan Illanes removes the paper clip–size bulb from the headlight of one of the vehicles, and uses some strands of telephone wire to connect it to a battery he's removed from one of the nineteen vehicles trapped with them. From then on, a weak, gray light hovers over the praying miners. To Yonni Barrios, in that light, they all seem to grow taller. He knows it's an illusion of light and shadow, but there's something magical about the way they look under that small bulb, standing or kneeling, listening to the word of God.

When Víctor Segovia begins to write again, it's to say he cried during the prayers led by José Henríquez. He addresses his daughters. "I deeply feel the pain I am causing you," Víctor writes. "I would give everything to soothe your pain, but it's not in my hands." He's absorbed the ideas of the Pastor's sermon, of his own smallness before the mountain and God's judgment. In the course of a single day of writing in his diary he reflects on his life—"Now I understand how wrong I was to drink so much"—and he gets closer to accepting the idea that he's going to die in the San José Mine. "Never in my life did I think I would die in such a manner," he writes. Just days earlier he'd been at home surrounded by the things that made him feel good: his music, his friends from the mine, gathered for a party at his home. "I don't know if I deserve this or not, but it is very cruel." He begins to say goodbye, to his daughters, his parents, his grandchildren, promising: "I love you and will look after you wherever I am." A few hours later he writes: "I feel guilty for causing you this pain. I should never have been in this mine knowing the shape it was in." He begins to give his daughter Maritza instructions for settling his affairs and begs her to help his mother with her debts. Maybe someone will find their bodies, eventually, and take this note to Maritza, and if that happens Víctor will have cared for his family, in some small way, from beyond the grave.

Someone says that if you heat up food it has more calories and more nourishment. So on the third day they're trapped the miners decide to cook a batch of soup and have a kind of picnic inside the mine, by the place where the mechanics used to work and the air circulates a bit. They manage to get all of the men out of the Refuge for the walk uphill to that place, on Level 135.

In the middle of a gray pile of stones, they make a fire that's about as big as two cupped hands. They remove the cover from the air filter on one of the big machines, turn it upside down, and make that their pot. José Henríquez has a cell phone he brought into the mine and realizes he can use it to record this event—but he doesn't know how to operate the camera, so he gives it to Claudio Acuña. Mario Sepúlveda becomes the main narrator of the video, speaking to Acuña and the camera with a voice that suggests he believes outsiders will find this recording one day. "Tuna with peas!" he announces. "Eight liters of water, one can of tuna, some peas. A little tiny fire here. So that we can survive this situation!" Around Sepúlveda, men move about with yellow and red helmets on, most of them shirtless, and a few are sitting on the pile of rocks by the fire, a dancing ball of orange light near the center of the dark frame of the video. Sometimes Acuña turns the camera and captures the lamp of one of the vehicles, but mostly the image is of a black space that's filled up with Sepúlveda's voice. "And we're going to show that we are Chileans of the heart. And we're going to have a delicious soup today," he says. Acuña turns off the cell-phone camera to save the battery, and a few minutes later he turns it back on to record Sepúlveda serving his completed soup to each man, using a metal cup that clanks against the bottom of the air-filter pot. He pours the hot liquid into plastic cups to several different men, and the water is murky colored.

"Has everybody got some?" Sepúlveda asks. "There's a bit more, if anyone wants it." He scrapes his tin cup against the bottom of the air filter and starts to talk to his son, as if he were watching the recording. "Francisco, when God tells you to be a warrior, these balls [*esta hueva*] are what it means to be a warrior." He imagines his son watching him, a warrior feeding these other warriors, refusing to surrender. *You see, Francisco, a warrior isn't just someone who slays dragons—or Englishmen, like Mel Gibson does in our favorite movie,* Braveheart. *A warrior can also be a man who takes apart an engine to make soup and then serves it to his brothers, keeping up their spirits with the rising inflections of his voice.*

Acuña stops recording so that he can pray with the others when Henríquez blesses the meal. As the men around Henríquez bow their heads, he gives thanks to God for the food they are about to eat. Then they sit on or near the pile of rocks and sip from their "soup," with its sheen of oil that might be from the tuna but also from the clouded industrial water they used to cook it. And sitting there, in that relaxed and convivial moment, a few of the men remember the last big meal they shared together, at Víctor Segovia's house in Copiapó, when Víctor had most of the men from the A shift over for a party, on a Wednesday afternoon after their last shift was over, before the men from the south departed for home on the overnight buses to Santiago.

It was two weeks earlier in a neighborhood where the streets are named for minerals. Víctor Segovia's house is on Chalcopyrite Street. Alex Vega brought a big pot that's called a *fondo* in Chile. They were going to make *cocimiento*, with chicken, pork, fish, and potatoes with the peel still on, and a little bed of cabbage at the bottom. The recipe says to cook all this in water, then pour in some wine near the end. They set it all to boil, and then they imbibed Cristal beer and some wine, but not Mario Sepúlveda, because he's a Jehovah's Witness and doesn't drink—he agreed to keep an eye on the pot instead. After a few drinks Edison Peña picked up a microphone—Víctor Segovia is a musician, he has a lot of sound equipment around the house—and started singing some Elvis tunes, including "Blue Suede Shoes," in his thick, Chilean-accented English. "Hey, that's music for old people," Pedro

Cortez and some of the younger guys said, teasing Edison, because they live in the world of reggaeton and cumbia, and that old music from the U.S. South sounds like something their parents might listen to.

Yeah, we were all having a pretty good time, out there at Víctor Segovia's house, the men remember. But then the party ended, not long after Pablo "the Cat" Rojas's cell phone sounded at about 4:30 p.m., just when Sepúlveda was saying the *cocimiento* was ready to go and the scent of stewed meats was floating about the house, and the guys were feeling nice and warm inside from the wine and the beer. Pablo Rojas and Víctor Segovia are cousins, and the call brought word that Pablo's father had just died. This was not unexpected news, because the elder Rojas was a lifelong miner who had become a hopelessly addicted drinker after he retired. You could find him in the plaza of Copiapó, after drinking for several days at a time, begging for money to drink a little more. José Rojas had been slowly killing himself for a few years, and now the inevitable had happened, and the news hit Pablo pretty hard. Pablo didn't cry, but his cousin Víctor could see he was in a bad way, so he said that maybe Pablo should go to the hospital to see his father and not worry about the party and the meal.

No one was in the mood to eat after Pablo left, and the party broke up pretty quickly—Luis Urzúa arrived late, when the last of the guests was leaving. Urzúa didn't have any food either, and that huge pot of *cocimiento* went uneaten.

"All that food! And we just left it all there and went home!" Pedro Cortez shouts out now, as they sit around a pile of rocks at Level 135. Freshly cooked pork, fish, and chicken, boiled in white wine, in a huge *fondo*. And now they're drinking a cup of "soup" cooked in a truck's air filter from a single can of tuna and eight liters of Mario Sepúlveda's bathwater, with no salt and only a few peas and some motor oil for flavoring—all divided among thirty-three men!

It's funny what can happen to a miner in the course of just two weeks: They've finished one shift, working hard to lift gold and copper from the earth, and they've lived well, cooking up big, miner-size portions of food to eat, with beer and wine to drink (even if they never did get around to eating the meal); and they've lost a father and an uncle who killed himself when he couldn't be a miner anymore; and then they've gone back to work, got trapped in a mine, made soup from water meant for machines, and blessed that dirty water as "food" and shared it with their brothers. If they manage to get out of here, they will tell this to the people they love, a story about food, family, and friendship. It's a tale of two meals: one aboveground, with nice dishes, lots of food, and very little eating; and one belowground, with very little food, in which they licked the insides of their plastic cups.

After the meal, a few of the men get excited because they say they hear the sound of distant drilling.

"I can feel something vibrating," someone says. "I hear it." Everyone turns very quiet, to see if they can hear it, too.

"It's a lie, you can't feel anything," someone answers, and the discussion goes back and forth for a while until, finally, even those who said they felt that faint and possibly imaginary vibration concede that it's stopped, or disappeared, or that maybe it never even existed. Víctor Segovia throws himself on the warm mud of the Ramp outside the Refuge, and fights off depression by writing in his diary. "Down here there is no day, only darkness and explosions." He describes how the men around him are sleeping, some using plastic soda bottles as pillows. Víctor and the others feel the monster of "insanity" welling up inside them, he writes. It's his fourth day underground now. He draws a diagram of two dozen prone stick figures, scattered before a doorway in the rock marked REFUGIO, and in its stark crudeness it resembles the grim police sketch of some crime scene. He lists the names of his five daughters again, and of his mother and father, and of himself, and then circles a heart around them. "Don't cry for me," he writes. "We had good times, always, with our *azados* [barbecues] and making *cocimiento*."

At the next prayer session, at noon, the Pastor tries to keep them strong and Víctor records his words. Being trapped in the San José "is a test that God set before us so that we can think about the things we've done in our lives that were wrong," the Pastor says. "If we get out of this place, it will be like being reborn into the world."

At 4:15 p.m. they think they hear drilling again. Two of the men get very excited and start to shout, but within an hour, the sound is gone.

Víctor has broken into a rash again, from the heat and from the worry. All the excitement of hearing the drill evaporates and Víctor studies the now-quiet men around him. "We look like cavemen, full of soot, and we are skinny, which is very noticeable on most of us."

Finally, at 7:15 p.m. on August 8, seventy-eight hours after being trapped, Víctor records the sound of something spinning, grinding, and hammering against rock. For three hours it grows steadily louder. At 10:00 p.m., Yonni Barrios is ready to believe it, too. It's unmistakably a drill, the sound traveling through two thousand feet of rock. Omar Reygadas says it's a dirt drill, because those are the ones that use a hammer: A diamond-tipped drill doesn't make that much noise. Soon it seems to be everywhere, coming from every wall. It grows louder, and to men who work with such tools, the pneumatic pressure at work is palpable. Rat-tat-tat. Grind-grind-grind. It's a drill, powered by air, working its way down the rock, and it's headed toward them, as made clear by the fact that as the hours pass the sound grows louder.

"Do you hear that, *huevones*?" Mario Sepúlveda shouts. "Do you hear that? What a beautiful noise!"

Someone up there is coming for them.

"Those drills can make one hundred meters a day," someone says.

Everyone starts doing the math. Maybe by Friday or Saturday, at the earliest, they'll break through, which means another five or six days without a true meal.

When the men eat their daily cookies, some allow each bite to stay in their mouths a long time without swallowing, because the taste itself is sort of like eating, as if they were ingesting an entire package of cookies, and not just two. Even just a few days of hunger can lead a man to do things he might not do otherwise, which explains why one day the saline solution that was in the first-aid box in the Refuge suddenly disappears. "The saline solution is gone, *niños*," someone says at the daily meeting. "Let the person who took it please step forward and return it. Or if you drank it already, let us know." No one steps forward, even though several of the men know who is responsible for its theft: Samuel "CD" Avalos, the vendor of pirated music. "I stayed quiet," Samuel later says with a chuckle. "It was just one of my crazy things I did one night." He's been secretly sipping at it, and has drunk about half the plastic bag already. "It tasted salty."

"Well, if no one knows where it is, we need to find it," someone says. "Everyone start looking."

The men make a show of looking for the precious bag of saline solution, including Samuel himself, who declares suddenly, and not without irony, "Oh, look, here it is. I found it!"

After that, Mario Sepúlveda pours a few salty drops from the bag into the glasses of water along with the spoonful of tuna the men get each day. And sometimes he inadvertently adds another salty ingredient to the water. Several of the miners notice that as Mario pours the water and puts a few peas or a few drops of milk in each cup, the sweat from his forehead drips into some of the glasses: He's telling the men how good everything is going to taste, and is simply too excited and too wrapped up in what he's doing to notice. Not only are the men eating meals made from Mario Sepúlveda's bathwater, they're drinking some of his sweat, too.

7

BLESSED AMONG WOMEN

The driller Eduardo Hurtado reaches the San José Mine on Sunday at 9:00 a.m., after an all-night drive of 430 miles, having been summoned by his boss at the Terraservice drilling company the night before, a few hours after Minister Golborne's tearful announcement that a "traditional rescue" was no longer possible. The machine Hurtado will use arrives two hours later: a Schramm T685WS rotary drill, manufactured in West Chester, Pennsylvania. It's a portable drilling rig on wheels, a vehicle about as long as a gasoline tanker, and it carries a mast that rises up to direct a hydraulic-driven drill into the ground. On an average day Hurtado and his crew will take this kind of drill and, with the guidance of a geologist or topographer, sink a hole that searches for ore. "Yeah, I'd done a lot of deep holes, but always looking for minerals," Hurtado says to me. "I'd never drilled for *viejos* before." Other drillers have already started searching for the men, the first beginning on Saturday night. Hurtado steps into the small company office at the mine and finds Alejandro Bohn, one of the owners, looking exhausted and dispirited, though his manager, Pedro Simunovic, is much more alert and helpful. Hurtado needs a topographer to tell him where to drill, but for the moment there isn't one to be found. "The situation felt chaotic and anarchic," Hurtado says. "No one was in charge."

Finally, Hurtado finds a topographer and some blueprints of the mine. It's obviously much faster and more accurate to drill a hole straight down, from a spot on the surface directly above the location they're trying to reach, in this case, one of the passageways near the Refuge. The hunt for this ideal drilling spot leads them to climb the bare, rocky surface of the mountain and mark out a patch of ground, which a work crew then begins to flatten with a bulldozer, to create the "platform" upon which they'll place the drill rig. They're not quite finished when a geologist inspecting the area tells them to stop: He's found several telltale cracks in the mountain. They are standing above the vacuum in the mountain left by the collapse of the mine. "This could all give way at any moment," the geologist says.

They hunt for another spot, where they'll have to drill diagonally into the mountain, and by late Monday morning they're ready to go. Hurtado feels the need to give a little speech, reminding his crew of eight men how important this job is, and says that maybe each guy should say a prayer for the hole they're about to drill: "Let's all put our trust in the skinny guy," Hurtado says, meaning, of course, the skinny guy on the cross. As they bow their heads one member of the drilling team says: "Hey, boss, let's hold hands as we pray." The eight drillers stand in a circle of helmets, and then the operator, Nelson Flores, places a rosary on the drill. Soon the rig's compressor and rotary drill hum and grind into action, the truck's mast tipped at a 78-degree angle and aimed at a target below 2,000 feet of diorite. "It was going to be hard because of the angle," Hurtado says. "I could end up anywhere. It's impossible to control exactly what the deviation will be." As they drill the shaft, the Terraservice crew will place a series of interlinking steel tubes in the hole they're carving into the mountain. Gravity will cause this steel shaft to bend in the same way a series of linked plastic straws bends. If they hit the Refuge, it will be with what soccer-loving Spanish-speakers call a *chanfle*, a bend-it-like-Beckham curve shot. The deviation can be as much as 5 percent, meaning that by the time it reaches the level of the trapped men the drill bit could be 100 feet off its target—and the corridor they're trying to reach is no more than 32 feet wide.

The grinding and pounding Terraservice drill spits a constant cloud of dust skyward from a chimney pipe, and sends a flow of wastewater over the ground. The sound and the dust fill the cold night as a fog begins to descend over the mountain. On other patches of mountain around him, more teams begin to drill, too; Sunday is drifting toward Monday, and for the moment, no one is in charge of coordinating the rescue effort.

•

When Cristián Barra arrives at the mine on Monday, the sense of disorder is palpable, and troubling, because it's his job to prevent chaos. Barra is there at the behest of President Sebastián Piñera—Barra is one of those strict and severe men who work behind the scenes to keep the best Latin American democracies running with a no-nonsense, kick-'em-in-the-balls-and-get-it-done efficiency. He works at the Ministry of the Interior, traditionally the most powerful agency in any Latin American country, and the one in charge of Chile's police forces and security apparatus. Barra seeks out the mine owners, Alejandro Bohn and Marcelo Kemeny. He finds them in their small office with the tiny window facing the desolate stretch of desert mountain range they own, two middle-aged, sleep-deprived men in oxford shirts and white helmets. They tell anyone who will listen that they thought the mine was safe—Kemeny says a few months back he entered the mine with his own two sons, ages fifteen and nine.

The night before, Kemeny and his manager, Pedro Simunovic, had a brief, tense encounter with the families. The angry men and women staged a protest to force the owners to "show their faces." Amid much pushing and shoving despite the police officers assigned to protect them, Kemeny and Simunovic entered a tented meeting area set up by the local government. Simunovic withstood a hail of insults and was able to utter only a few words, while Kemeny stood in the background saying nothing, and most of the family members didn't even notice he was there.

The rescue efforts are moving forward without the owners' input. "They weren't psychologically or emotionally able to make any decisions, or to plan what to do next," Barra later says. Anticipating this state of affairs, Barra has come armed with an official declaration of a State of Emergency, issued by the minister of the interior at the order of the president, at a meeting at La Moneda, the presidential palace, a few hours earlier. Barra is Piñera's fixer; they've known each other for more than twenty years, since the early days of the National Renewal party. Barra has come to take charge of the mine, and one of his first acts will be to deploy police officers and erect barriers that will keep the miners' fathers and brothers and sons from entering the mine on more quixotic rescue attempts.

Barra also establishes a series of protocols—who can enter the mine property and who can't, the identification required to pass through. He is the vanguard of an army of federal officials on their way not just to rescue the thirty-three men trapped beneath their feet but also, in a sense, to rescue the minister of mining. Like everyone else in Chile the top members of the Piñera government saw the minister cry on television because he couldn't tell the miners' families exactly how he would get the men out alive. Now the Piñera administration has assumed the responsibility of giving him a plan, despite some grumbling from the president's advisers that it's not politically expedient: Why assume responsibility for the lives of thirty-three men who are probably doomed anyway, when tradition and the law dictate that you need not and should not get involved?

The president has made a quick stop in Copiapó on his return journey from Quito, Ecuador. He met briefly with a small group of family members, and with several local officials, including the provincial governor, Ximena Matas (she is one of his appointees), and a pair of conservative members of Congress, though not with the leftist legislators present, including the Socialist senator Isabel Allende, a cousin to the novelist of the same name. Later, in Santiago, he convened that first meeting at La Moneda at which he agreed the government should take over the rescue. The next obvious question is: Who knows more about mining rescues than anyone in Chile? It has to be someone at Codelco, the state-run National Copper Corporation, the world's largest producer of copper. Soon a call is out to the man who runs the largest of the Codelco family of mines, André Sougarret, at El Teniente mine in Rancagua, south of Santiago. Sougarret is an engineer and administrator at a mine that's so big—it has seven thousand employees—that rescues are a routine part of the worklife there. As he begins to assemble a team of about twenty-five men he gets a second call with an urgent question: How quickly can you make it to the presidential palace?

Ninety minutes later, Sougarret is entering La Moneda, Chile's equivalent of the White House, for the first time in his life, dressed in jeans and carrying a mining helmet under his arm. He's directed to an office where he waits for a meeting that never takes place. Two hours later, an official tells him to go down to the basement. "I had no idea what was happening," he later says. Finally, he's told: "You're going to Copiapó." He becomes a passenger in a caravan of cars headed to the air force base adjacent to Santiago's international airport. When he arrives at the base, Sougarret sees, to his surprise, that he's getting on a plane with President Piñera. A bit after takeoff, a crew member serves Sougarret lunch, and when he's finished eating the president and the first lady emerge from their private cabin and sit next to him. The president pulls out a notepad and sketches a drawing that shows what he, the president, knows about the San José Mine and where the men are trapped. The president says something to the effect of "Well, that's the situation. If I give you absolutely any resource you need to rescue them, what's the probability of getting them out alive?"

Sougarret can't answer, and neither can another engineer sitting next to him. "Then he asked us if we knew of any other kinds of rescue, something that might work in this situation," Sougarret says. "We told him that, in general, you can't really predict if a rescue will work—and that, generally speaking, there are more negative outcomes than positive ones."

When the presidential plane arrives in Copiapó a little after 4:00 p.m., it's very cold. Sougarret gets into the backseat of a van that takes the president to the mine, and there they join Golborne for a brief press conference in which the president announces the government has brought in the country's top mining-rescue expert to lead the effort to find the trapped thirty-three men—he then names Sougarret, though he mispronounces his name. To his great relief, Sougarret manages to escape the press before they can ask him what, exactly, he plans to do.

One of the first men Sougarret meets at the San José Mine is the general manager, Carlos Pinilla. "Hey, remember me?" Pinilla says. "I met you at La Serena." Decades earlier, Sougarret was an intern at that mine when Pinilla was the boss there. Pinilla and the other managers provide Sougarret with information that gives him some hope. He learns that there are likely several thousand gallons of water stored in tanks inside the mine, which means the trapped men, if they're alive, won't immediately die of dehydration. The San José is more than a century old and thus has many forgotten passageways that allow air to seep in and out. In fact, as he stands on the Ramp near the entrance, Sougarret can feel air flowing into the mountain: Any men trapped inside will likely not die from suffocation. Going deeper, Sougarret also verifies that "it was a very good mine, in terms of the rock that was holding it together." This is at once reassuring and disturbing: The hard diorite shouldn't have caved in, but it did, which means that the essential structure of the mountain must have failed. Whatever is blocking the many passageways leading to the men must be a very large obstacle indeed, as is soon verified by a group of geologists who estimate the skyscraper-size "mega-block." It would take a year to excavate a new tunnel around that obstacle to reach the men.

From the medical personnel who have arrived at the mine, Sougarret learns that a healthy man can last thirty to forty days without food. However, if there's a man debilitated with a lung disease such as silicosis (and they'll soon find out there is, as Chilean health officials round up the medical records of the men), he'll survive perhaps half as long; and a man with a broken limb or some other serious injury might survive as little as a week or two. Four days have already passed. They have to try every possible rescue strategy, and Sougarret decides to send one team to fortify the passageways below so that a second team can try to reach the men by clearing the chimneys. Many among the dozen or so mining professionals who have come to the site to offer guidance and expertise believe that this sort of "traditional" rescue through the chimneys and other passageways is the best hope.

The "nontraditional" effort consists of nine drills working independently—in effect, the rescuers are firing nine bullets at the same target and hoping one will hit. Hurtado, like the other drillers, knows all of Chile is watching. After three days of drilling, the Terraservice borehole reaches a depth of 370 meters: Hurtado's team stops and pulls out the drill so that a topographer can use instruments to check on its progress. The report back is not good: The hole has bent in the wrong direction. Can he make it bend back the right way, someone asks? "Impossible," he answers. "It was as if we had set off for Caldera, but ended up on the road to Vallenar instead," Hurtado later says, naming two towns that are on opposite ends of the Pan-American Highway as it runs through

Copiapó. Among his drill team, one man looks especially beaten down by this setback: the man who called them all together to hold hands at the start. "Our feelings were heavy," Hurtado says. "This wasn't an ordinary hole anymore." They begin to drill again, the geologist Sandra Jara checking their progress every 200 meters with a device that Hurtado's team lowers into the 4.5-inch-diameter borehole. It contains a gyroscope that uses the Earth's rotation and some basic principles of physics to establish true north. This hole, unlike the first one, seems to be bending in the right direction, and teams working in six more twelve-hour shifts push it down to 400 meters, then to 500, working with a combination of urgency, altruism, and pessimism, because beyond the likelihood that they will miss their target, there is also the distinct possibility that if they do find anyone that person will be dead. The possibility of finding corpses is real enough that Barra and the Ministry of the Interior have put a special protocol in place in case the drill breaks through: Sougarret will supervise a team that will lower a camera down the hole, but only he, the minister of mining, and the camera operator will be allowed to see the monitor, because it might reveal the macabre image of a dead man, or several dead men, or even thirty-three dead men. If the men are dead it will fall to the minister, and only the minister, to tell the families.

The drilling proceeds for a fourth, fifth, and sixth day, each night ending for Hurtado as he travels to Copiapó to sleep. One day he's driving back and he spots a woman with coffee-colored skin and indigenous features standing at the crossroads that links Route C-351 to the spur that leads to the mine. She looks not much older than a teenager, and as she turns to face his pickup truck she sticks out her thumb: She's one of the miners' relatives, hitchhiking to Camp Esperanza. Hurtado isn't supposed to talk to the families—another of the many rules Barra and his team have established—but he stops to give her a ride anyway. She introduces herself as Veronica Quispe, the wife of the trapped Bolivian miner Carlos Mamani. They exchange a few pleasantries and she mentions how the water was recently cut off to her Copiapó neighborhood, a common-enough indignity in the type of informal *campamento* settlement where Bolivian immigrants and other poor people live. He leaves her at the gate. In the days to come he sees her several times more, waving as he passes through the gate: She's sitting under an umbrella and doesn't have a tent like many of the other women do. Then he returns to the drill and wonders if it really is bending in the direction of Veronica Quispe's husband.

A week after the accident, President Sebastián Piñera announces the resignation of the head of Chile's mining regulatory agency, the National Geology and Mining Service, and two other top agency officials, for their failure to monitor the San José Mine. The rescue effort is a potential public relations disaster for the Piñera government, and also a potential bonanza. The president's advisers, mindful of their responsibility to lay out the risks and benefits of any decision to their boss, conduct a public opinion poll about the personalities linked to the disaster and rescue, according to Carlos Vergara Ehrenberg's account of the disaster and rescue, *Operación San Lorenzo*. Piñera and Golborne score very well, while the mine's owners have become among the most unpopular people in Chile.

The mine owners, Bohn and Kemeny, live in Santiago, but they return to Copiapó on August 12 to grant interviews to two of Chile's biggest newspapers. They don't take any responsibility for the accident and suggest the miners themselves might be to blame. Even worse, from the Chilean government's point of view, they dress in the same red jackets as the government officials working at the rescue site. "Those idiots screwed us," says the president's fixer, Barra. "Why did they have to wear those red jackets?" A few days later, Golborne commits a much more serious faux pas: He tells a television interviewer that he believes there's little chance the men are still alive. Golborne quickly retracts this statement, but the truth is the government is already preparing, privately, for the worst, according to Vergara Ehrenberg. If the men can't be found, the government will seal off the San José and declare it a "sacred place" that can never be mined again.

When men were still extracting ore from the San José, women were told to stay away from the mine. A woman in a mine is said to be bad luck, but now a growing group of girlfriends, daughters, sisters, and wives occupy the entrance to the San Esteban Mining Company property. Watching the first rays of sun burn through the fog to illuminate that windswept patch of ground is an eye-opening experience for many of these women, especially those such as Carola Bustos who have come from Santiago and other cities to the south. In the light of day it is immediately apparent how far removed the mine is from civilization. There is the sadness of the narrow, crumbling asphalt strip that leads from the main highway to the front gate: Driving it feels like watching the opening scene of a movie about a remote and hopeless place. They see how small and dilapidated the mine company buildings are; and when they look at their cell phones they see their husbands and boyfriends were speaking the truth when they said there's no signal here. Some of the women are angry with their men for taking a job at such a remote, forgotten, and patently dangerous place—because now they can see and feel how crude and simple the mine is, how it's just this hole carved into a mountain and not the safe and orderly work site they imagined or hoped it was. Their men said it paid well, after all, better than any other job they could get, but evidently little of the enormous wealth created by this mine was put back into it. All these men working here: They were just looting this mountain, weren't they? Rushing in to get out as much gold, literally, as they could, before the very stone into which they were tunneling came tumbling down upon them. Some of the wives and girlfriends are angry with themselves for allowing themselves to be fooled, despite the now-obvious clues that the tough guy who slept next to them was trying to give about what the San José was really like. But now they're here, once again forced to keep home and family together after men have messed everything up. Of course, it isn't their man who's to blame for this particular mess, but rather the men who own the mine. It's up to each woman to fight for her man, because his kids, their kids, need him back home, with all his hard-drinking, wandering-eye flaws, and his *mal genio* moodiness.

The mothers, especially, feel their instincts and emotions pulling them in different directions. When Mario Sepúlveda's wife, Elvira Valdivia, arrives in Copiapó, she travels immediately to the mine with her children, Francisco, age twelve, and Scarlette, age eighteen. But after a short while amid the weeping wives and the shouting mothers at the mine entrance, and the mine officials and the Carabineros yelling at the women to get away from the front gate, Elvira realizes she can't spend the night there with her children. Scarlette is looking more like a girl than a woman suddenly, she's relapsed into a state of frightened helplessness, and Elvira speaks to a doctor who gets her a prescription to help Scarlette sleep. Elvira also forces the company to pay for a hotel room facing the plaza in downtown Copiapó. At night, in that room with Francisco and Scarlette, she reaches deep into her feminine soul and calls upon the strength she needs to protect them from even the idea of death and loss. A mother's instinct tells you your children need you to make them believe they will see their father again, but this isn't an easy thing to do when you've seen the rocky mountain that's swallowed him up.

Still, for those first days, the notion that Mario isn't like most other men makes it easier to believe he'll come out of the mine. "I knew that Mario wasn't going to let himself die," Elvira later says. "No, Mario is the type of person who will eat someone to survive if it comes to that. And if he had to eat mud to survive, he'd eat that, too." For these first few days Elvira is certain Mario is alive. "If I had thought there wasn't any hope I would have turned around and left and gone home because I'm cold-blooded when it comes to those things." She worries, however, about his not taking the medication that keeps his emotional pendulum from swinging too wildly. Elvira has spent two decades enduring the mood shifts of the man with "the heart of a dog" and the series of never-permanent jobs that took him from one end of Chile to the other. The lesson of all those years is that Mario always comes back, he pulls himself out of unemployment or depression and finds a way back home to make Scarlette laugh, to be Francisco's hero. Every night, she gathers her children in the hotel room to pray.

Mónica Avalos, on the other hand, plants herself at the mine entrance and doesn't leave. For the first few days and nights she barely sleeps at all. Mónica is, by her own account, falling apart: She's even lost track of her seven-year-old son, Bayron. She's at the mine when she realizes she doesn't have him—her husband's friend Isaías tells her that his wife is taking care of the boy. "Don't worry, *la* Nati has him. He hasn't got any clothes, but they're going to get some for him." Mónica's teenage son, César Alexis, tells her she needs to go home and bathe and sleep, but she won't. "I didn't care about taking a bath, I didn't care about eating. I didn't care about anything. Nothing. Nothing. People would tell me, 'Mónica, you have to be strong,' but no, no, I couldn't be." She spends her first night at the mine sleeping for a few minutes while sitting on a slate-colored rock; the second night she finds a wooden pallet, and curls up on that for a short while, still dressed in the sweatpants she was wearing when she was making Florencio the soup he never ate. Another night she's awake standing on the mountain under the stars; she sits down on one of the brick-shaped blocks of stone, closes her eyes, and falls asleep, and when she opens her eyes again she's awake at another spot far away. She's been sleepwalking over the mountain where her husband is trapped, a *sonámbula* whose subconscious is pulling her over the gritty surface of the ground, one step after another.

Her teenage son, César, meanwhile, is channeling Florencio and trying to be brave and responsible: Among other things, he's going to school every day while also traveling to and from the mine. He's in the third year of high school; at 5:00 p.m. he gets out of class—a little earlier sometimes, if he asks for permission—and gets a ride to the mine, sometimes on the bus the Copiapó city government has set up for the family members. If he misses the bus, he hitchhikes. At his school no one knows for the first few days that César's father is trapped, but then the school's administrators find out. Take a month off school if you want, they say. "But I didn't want to be absent, because I was going to fall behind in class and my exams might not go well." So César Alexis Avalos, son of the trapped foreman Florencio Avalos, goes to class all day. When the final bell sounds, he heads to the mine to check on his mother and see if there's any news, and then he hitchhikes back to Copiapó if he has to, so that he can be at school the next morning. "The only thing my father wanted," he later explains, "was for me to go to school." César plans to honor his father's wishes, and to stick to the rituals of responsibility and hard work his parents built into each day, perhaps because doing so is a kind of unspoken acknowledgment that his father is still alive.

At this point, it takes a certain stubbornness to believe the thirty-three men are alive and that they will one day walk, or be carried, out of the mine. A newspaper in Santiago has put the chances of their success-ful rescue at less than 2 percent. Other media outlets report that the men can survive a mere seventy-two hours underground, and now they've spent nearly twice as long trapped. Carmen, the articulate, poetry-writing, and deeply religious wife of the shift supervisor, Luis Urzúa, has heard people declare her husband dead already. "The *jefe de turno*, he was with Mr. Lobos, in the personnel truck, and they were coming out when the mine collapsed," someone tells her not long after she arrives at the mine. "The truck was crushed. They're both dead." Carmen refuses to believe this and snaps back: "If he was dead they would have brought out his body already through the chimney!" But it's deeply unsettling, because the verbs and the adjectives of death are there, hovering over the mine entrance. "*Están muertos.*" They're dead. "*Murierion.*" They died. Officials at the hospital have repeated these words, too: "They're dead," one told Alex Vega's wife, Jessica, causing her to faint. The fear of death is threatening to consume the women around Carmen, women like the sleepwalking Mónica Avalos, with her swollen eyes and frizzy hair. No, sisters, don't believe what they're saying, Carmen says. Hold on to your faith. "We need to pray." Carmen teaches catechism classes at her church, and she takes her silver rosary—"I always carry it with me"—and begins to pray on a cold night, forming a circle with a few other women. A few days later she spots a small plaster statue of the Virgin of Candelaria: a replica of the one that resides in a chapel in Copiapó, it-self found in the eighteenth century in the nearby Andes, the figure of a woman said to have appeared miraculously in a palm-size stone. These plaster figures are ubiquitous in the mining north, and now Carmen and the other women decide to build a shrine in Camp Esperanza with this humble representation of the mother of Christ at its center. By now the local government has put up a tented field kitchen to feed the relatives gathered in the camp; the women build their shrine nearby, next to the spot where the government passes out bread. They place a few stones around the plaster figurine, then put the statue inside a cardboard box so that the wind won't blow out their votive candles. "We made a small place where people could go and let go of their pain, where they could pray for the miners, and start to forget that they might be dead," Carmen says. They kneel before the image of the Virgin Mary, in thick wool sweaters and big parkas, underneath winter caps and hats, and they recite the prayers of the Rosary, an Apostles' Creed followed by repetitions of the Our Father and the Hail Mary, its "blessed art thou among women" whispered in chorus.

More shrines arise on the mountain, many built to individual miners on the scree of stones, with candles affixed to the rocks with dripping wax. Prayer seems to be their only defense against the growing sense of hopelessness and finality. André Sougarret has ordered the rescue team he sent to fortify a route into the mine to stop working. By spray-painting marks on the surface of the gray guillotine of stone blocking the Ramp, they've detected that the vast, destructive "mega-block" at the heart of the mine is still moving. The broken skyscraper of stone inside the mountain is slipping downward: A new collapse is possible at any moment. That night Golborne and Sougarret hold a press conference announcing the closure of the mine. No one will be allowed in, the entrance will be permanently sealed. After the difficult, emotional encounter with the families that follows, Sougarret returns to his hotel. He falls asleep at midnight, but fifteen minutes later a call wakes him up. Eduardo Hurtado and the drilling team from Terraservice have hit an open space with their second borehole, at a spot 504 meters below the surface, but almost 200 meters above the Refuge.

As Sougarret travels to the mine, word spreads of the Terraservice discovery, and all the drills stop working, because the drillers need the silence to listen to their shaft. It's the night of Sunday and Monday, August 15–16, ten days after the men were trapped. The Terraservice men place their ears to the uppermost of the steel pipes they've lowered into the shaft: They hear a rhythmic noise, a tapping. Hurtado asks one of the police officers to listen: "Can you hear it?" The officer says he can. A short while later Sougarret arrives and places his ear to the metal. He's not sure that what he hears is a sound made by humans. At 1:00 a.m. the crew begins to lower a camera into the borehole. It is a night of especially thick fog and wind, an ominous sign to Hurtado. At 6:00 a.m. the camera finally reaches the bottom. As per the protocols of the Ministry of the Interior, Sougarret and the operator are among the handful of men allowed to see the monitor. But soon Sougarret is telling the Terraservice drillers what he sees, and allowing them to look for themselves. There is nothing. Just a space of empty rock, a cavern, or *socavón*, that appears to have been excavated and emptied of ore. As for the tapping sound? "The power of suggestion," Sougarret says. "They wanted someone to be down there, so they were hearing things that weren't there."

The days pass, and the pessimism grows, threatening to envelop even the most lively and hopeful and determined of the miners' wives and girlfriends. Susana Valenzuela, the other woman who lives with Yonni Barrios, hears some of his estranged relatives at Camp Esperanza say he's probably dead already. Susana has been going to the mine with Marta, Yonni's wife, and it's quite awkward, because Marta's family is there, too, including the adult children Marta had before she met Yonni.

Later, at home, Susana serves her boyfriend's wife a cup of tea as Marta talks about the news from the mine, which is very bad. "Listen, Susi," Marta begins. "Look, this is as far as you go. I've come to let you know. Because Yonni is dead. So I need you to give me Yonni's things." Marta wants Yonni's documents, his pay stubs especially, which she'll need to receive his death benefits from the mine owners and the government. Susana listens to this and thinks that Yonni isn't dead, and that this woman is sort of crazy, and if she wants his papers, why not? Those material things aren't important to her, she's worked her whole life and has her own savings and pension, so who cares? But as she leaves to climb the stairs to their bedroom, the one she shares with Yonni, she stops.

"Show me Yonni's body and I'll give you those things," she says. "Prove to me that he's dead."

"You're really stupid," Marta replies, in Susana's account, because Marta will deny this conversation ever took place. "How am I going to get him out of the mine? Am I Superman?" Yonni is, after all, buried under a mountain of stone as far as anyone knows.

"If he's dead, go ahead and keep your money, but hand over his body to me so that I can have a wake for him," Susana answers.

"You'll have to talk to his sisters about that," Marta says, in Susana's account of this moment.

"Get out of here," Susana says. "Leave, and don't come back to bother me."

When Susana goes to the mine, she hears people say the miners are all dead. She hears one of Yonni's relatives talk about collecting his death benefits, too. This is the way she remembers those dark days: Yonni's relatives treat her like a nobody, because her bond to that flawed, womanizing, and well-paid man is now broken and meaningless. Perhaps Yonni's relatives see a kind of justice being done: Susana was able to manipulate Yonni and his need to be loved to claim his paycheck when he was alive, but if he's dead she won't be able to take from him any longer. But Susana feels that Yonni is alive. "He was down there, fighting for his life. He wasn't handsome, but . . ." she will say later, her voice breaking off as she begins to cry. "He was fighting. I could see him down there, buried, buried in mud." She imagines him being swallowed up by that same gray, gritty slime that she cleaned off his boots and his clothes when he came home.

"They're all dead," she hears someone say again, and she goes back home and throws herself on the floor and begins to weep. "I felt like dying." Then she hears a voice. "Chana," it says. That's what Yonni called her, "Chana." "I swear it, to the little mother who's in heaven, I swear I heard it," she says. Her love for the missing Yonni Barrios is so strong, she projects it into the sounds and objects inside the house they shared, seeing a series of "miracles" that fill her eyes with wonder, because she believes that each one has placed her in the presence of the Holy Ghost. She feels the entire house tremble, but when she asks her neighbors if there was an earthquake, they say no. The flames on the votive candles she places at a small shrine for Yonni go out and then start burning again all on their own. One day she comes home from the mine and sees police gathered outside her home. The neighbors say that they called the Carabineros because there were noises inside Susana's house, as if someone were taking the house apart—they suspected that Susana was being robbed. When Susana opens the door and goes inside, however, she finds everything in its place. She's convinced that the noises were Yonni's spirit sending her a message: *I'm alive, Chana! I'm fighting, don't forget me!* Some of these things she tells to the psychologist who comes to visit her later, but others she won't: "Because I could tell he thought I was crazy."

Others will also report odd and seemingly paranormal occurrences. A miner's cell phone calls home even though he's still trapped with that phone deep underground. Others report seeing the spirits of the thirty-three men wandering around northern Chile. In a section of the Juan

Pablo II neighborhood of Copiapó, where many Bolivian immigrants live, a neighbor of the miner Carlos Mamani says she sees him in her front yard one night. In the traditional beliefs of Bolivia's Ayamara people, the spirit of a man who is near death will walk at night in the days before his death. The neighbor tells Mamani's wife, Veronica Quispe, and Veronica's mother that she saw Carlos sitting on the patio in front of her house one night. Carlos was wearing a cap, looking off to the side, the neighbor says, but when she approached him to talk he disappeared. Veronica and her mother get very angry at the neighbor who tells this story, because to an Ayamara ear it can only mean one thing: Carlos is dying. Why are you saying that, they tell her. Stop sharing your ugly stories with us!

The miracles and the visions of death and suffering are spreading. María Segovia imagines her brother putting on a brave face below. Quiet Darío Segovia, with his square face and stoic disposition, and the rugged countenance of an indigenous warrior. Once he was a little boy who needed his loud big sister to stand up for him. From her tent in the camp, she can hear the machines drilling for Darío and his thirty-two colleagues. There are now many more families around her, growing circles of brothers, cousins, from cities near and far away: from the opposite ends of Chile, and from nearby Vallenar and Caldera, towns where miners' families know what it is to wait while men dig in search of other men. There are more than one thousand people at Camp Esperanza now, so many praying people "it was like Jersusalem," Jessica Chilla, Darío's girlfriend and life partner, will remember.

María Segovia often climbs above this growing city in the desert to look for and listen to the drills that are searching for her brother. She learns the noise patterns of each rig, the habits of the drill crews: what a diamond-tipped drill sounds like, when they stop to change shifts. At night each drill rig occupies a pool of light in the ink-black cloud of total desert darkness. The noise and the light and the energy of the drill crews is reassuring. But then, after a few days, one of the drills stops, and then another. For several hours she hears nothing. María hikes again to the nearby bare, rocky mountain that looms over the mine site, and begins to climb. Looking down from the top, she confirms: There is no dust coming from the machines. She climbs back down the hill to the camp to let the other families know, and several women rush up to the gate by the guard shack. "You've stopped drilling! You stopped!" The women hit the pots and pans they've brought to cook meals, making as much noise as they can. Finally, Golborne comes to the gate to say, No, they've just broken one of the diamond drills, we just need to replace it. Please, be patient. In the days to come, María Segovia climbs the mountain more than once to check on the drillers and Golborne's promises. She's a grandmother of fifty-two, climbing this hill as if she were a Girl Scout, with other women following after her, scratching their way up the slope. "We became like wild pumas running over the mountain," María says.

Finally, Cristián Barra's police supervisors post Carabineros at the base of the hill to keep the women from climbing it. But the next time María Segovia decides she needs to see the drillers working for herself, the officer who sees her coming looks the other way. She reaches her post on the mountaintop, sees the masts of drill rigs standing nearly perpendicular, the clouds of diorite turned to dust rising into the air.

She takes precarious, sliding steps back down the mountain, to the camp of tents and tarpaulin where the families of three miners are camping together. At the Segovia-Rojas camp, where the families of the miner-cousins Pablo Rojas, Darío Segovia, and Esteban Rojas wait and sometimes sleep, the nights do not pass quietly. They sing songs under the southern stars or in a swirling fog, and sometimes a chant of "*¡Chi-chi-chi, le-le-le, mineros de Chile!*" is taken up. And sometimes stories are told of three cousins, all in their forties, who grew up in the valleys nearby when there were still days when water could be found flowing through the river.

8

A FLICKERING FLAME

For the first few hours the sound of those drills coming toward them is at once calming and exhilarating. Víctor Segovia can't sleep, listening late into the night and into the next morning. At 4:00 a.m on Monday, August 9, more than eight hours after he began to hear the drilling, he drifts into a dream. He's home, asleep in his own bed, and he hears his daughter calling to him: For a moment Víctor is in a bright and open space, free of the torment of the mine, until he opens his eyes and finds himself on the floor of the Ramp near the Refuge, lying on cardboard, and he's swallowed up again by fear and longing. At least now there are two separate drills headed toward them. A few hours later he takes note in his diary of the lightening mood around him: "We are more relaxed," he writes. "Down here we're all going to be family. We're brothers and friends because this isn't the kind of thing that can happen to you twice." All thirty-three men attend the daily prayer session, followed by the meal of the day: today a single cookie and perhaps a spoonful of tuna, or an ounce or two of condensed milk mixed with water. Afterward, for the first time, someone mentions the need to sue the mine owners for making them suffer like this. It's a subject that will come up again and again in the days to come. Juan Illanes, a well-read mechanic from the south, has suggested that if they're rescued, they should keep a "pact of silence" about the accident, telling only their lawyers what happened so that they have a better chance of punishing the mine owners in court. Esteban Rojas, a forty-four-year-old explosives expert, reacts angrily.

"What's the use of talking about money and lawyers while we're still trapped down here," he says. "That's crazy."

It's madness to think about your problems in the outside world when you're still buried and half-dead. "The drilling is going really slowly," Víctor Segovia writes in his diary a few hours later. "God, when are you going to end this torment? I want to be strong but I have nothing left to give."

Omar Reygadas notes that the air seems to be growing heavier and hotter. Before, the air in the passageway next to the Refuge seemed to be moving, but now it is still, making it harder for him to breathe. He has long white bangs that fall over his forehead, and that have the odd effect of making him look youthful and old at the same time. Now he's starting to feel his fifty-six years. "*Estoy mal, estoy mal,*" he says. I'm sick, I can't breathe. Is it his imagination or is the air not flowing anymore, he asks another of the older men, Franklin Lobos. Franklin has his own problems: He's elevating his knee, injured ages ago during his professional soccer career, and has wrapped it in a rubber mat taken from the floor of one of the pickup trucks. Humidity causes his knee to ache, and in the past few days a stream of water has started to run past his sleeping area, transforming everything to mud. "I need to keep this as dry as possible," he explains to the men around him. Franklin hears Omar's question and answers, Yes, the air is heavier, it seems to be circulating less than it was before. Maybe one of the hidden, open passageways in the mine was sealed up by one of the constant rockfalls they've been hearing. Omar takes a few breaths from one of the two oxygen tanks in the Refuge, but it doesn't seem to be working. Mario Gómez, the sixty-three-year-old miner with the missing fingers, has been using this oxygen because his silicosis-weakened lungs are struggling against a lifetime of damage done by working in passageways like this one; and he's been further weakened by a daily diet now less than one hundred calories.

The drilling continues unabated into the next day, Tuesday, August 10. At noon, their midday prayer ends with a recognition that it's the Day of the Miner, a national holiday. The Day of the Miner falls on Saint Lawrence's day, because he is the patron saint of miners according to a Catholic tradition that goes back a millennium. In Chile it's a day when mining owners pay tribute to their workers with a big feast for their employees and their employees' families. There will be no feast today, but they do take a moment to say a few words in honor of themselves and their labor, and to stop and think about how proud they should be to be miners. Chile was built on the labor of men who risked their lives and suffered inside mountains, and mining is tied up with Chile's national identity: Pablo Neruda wrote poems to the miners of the north, and Chilean students still grow up reading books such as Baldomero Lillo's *Sub Terra*, a collection of early twentieth-century stories about mining work. The men of the San José are miners going hungry inside a mine on the Day of the Miner, and the feelings of pride-tinged suffering this simple truth brings lead them to end their talk by singing the national anthem.

Víctor Segovia is moved by the sound of all thirty-three hungry men joining their voices together. "For that moment I forgot that I was trapped in a mine," he writes in his diary, but the sense of being his free and ordinary self is only fleeting. As the hours pass, the sound of drilling strengthens and then fades inside the rock, and it's hard to tell exactly where it's coming from. It begins to disappear into the stone. Where is it going? Is it still headed toward us? A couple of other miners join Mario Gómez in taking blocks of wood and objects and pressing them against the cavern walls to listen, trying to pinpoint where the drill sounds are coming from. As the possibility that the drills might not reach him starts to feel more real, Víctor begins to reflect on his life again. He's never traveled beyond the valleys around Copiapó, but he's rich in family, and a growing circle of relatives takes residence in his sorrowful thoughts. In his notebook he makes a long list of in-laws, cousins, and uncles, thirty-five people in all, including a few estranged members of his family he hasn't spoken to in years, asking forgiveness of the reader if he's forgotten anyone, because at this moment, "I really only have half a mind."

When the drilling isn't so loud, and when the men stop talking, Víctor and the others in and near the Refuge hear an intermittent rumbling sound. It doesn't come from the walls, or from a distant rockfall, but from inside the Refuge itself, and it's loud enough that Víctor takes note of it in his diary. Víctor doesn't know it, but this noise has a scientific name— borborygmus, the noises caused by the layers of smooth muscle inside the stomachs and small intestines of the men squeezing and pushing down food that isn't there. It's a gurgling rumble set off by the remnants of the very little food they put in their mouths a few hours earlier, a noise made louder by the echo chamber of an empty stomach. Each contraction is amplified and transmitted for other hungry men to hear, causing them to think about food a little more than they are already.

On a table in the Refuge, some of the men play checkers, from a set made with pieces of cardboard. Later, Luis Urzúa, concerned about the men in the Refuge beginning to wallow and snap at one another, makes a set of dominoes by pulling apart and cutting up the white plastic frame of the reflective traffic-hazard triangle in his truck. Higher up the Ramp, at Level 105, where the mechanics and Luis himself sleep, Juan Illanes is working hard to keep up the morale of the *viejos*. He tells them stories. Illanes has a deep baritone voice, and the clear, confident enunciation of a television anchor, and he's articulate, educated, and has traveled widely enough in Chile that he has many interesting things to say.

Mostly, Illanes, on the sixth, seventh, and eighth days of their involuntary fast, talks about food. "Have you ever seen a lamb being cooked? On a spit, over a fire?" he asks the men grouped around him, sitting on their makeshift "beds" of cardboard and canvas scraps over on the Ramp by Luis Urzúa's pickup. Several of the men say yes, they have seen a lamb on a spit. "Ah, but what about six being cooked all at once?" It might seem torture to speak of food to men who don't have any, but no one tells Illanes to stop as he continues with his cheerful description of how it was he came to attend such a banquet. "I was on the pampa. By Puerto Natales," he tells the miners. He was in the army during the near-war with Argentina in 1978. "I was out there, with fifty reservists, about twelve hundred meters, no, make that eight hundred meters from the border." It was Christmastime, a season of traditional feasts, and "all we had to eat was army food," the tasteless provisions doled out to men on the front. One of the soldiers, a local boy, said: "This isn't the way we do Christmas down here. We make big meals." At that point, another soldier spotted some horses nearby. "They were Argentine horses: big heads, all mangy, truly ugly animals," Illanes recounts with a chuckle. "I can do a little business with these," said the local soldier, who had the look of a gaucho, and he disappeared into the night leading several away.

"The next morning, we wake up, and there are *twelve* lambs on two poles, all skinned and cleaned and perfect," Illanes tells his fellow miners, several of whom are now grinning. Six lambs each on a pair of long metal rods hanging between posts. "So we all go to get firewood," scrambling around the treeless pampa in search of bits of scrub branches, he says. "Pretty soon, we had a nice little bonfire, a real tower of embers going. *Chiquillos*, it was beautiful." Illanes hears little sighs of comfort from some of the miners around him, who are no doubt imagining the sound of meat and fat crackling over a fire, but he's not finished yet. Because then, he continues, one of the gaucho soldiers showed up with a bag, and passed out a bit of golden tobacco to each soldier, and a piece of paper, and they each rolled a cigarette in the peasant style. "In short, *chiquillos*, it was a Christmas to remember."

Illanes recounts the story with such detail that it must be true, and in the dim light, and in his slow telling, the miners around him feel as if they've been listening to an old radio show. He tells them another military story from the south, of riding across the Chilean pampa on horseback, and his encounter with a fungus known as the *dihueñe*. To the northerners from drier climes who've never seen these delicacies, he describes them: "They're mushrooms that grow on the branches of trees, and especially roble trees when they're young." They're orange and honeycombed, the size of a walnut, with a sweet, clear liquid inside. "So there I am horseback riding, when I come upon this shrub that's not six feet high. And its branches are all covered, top to bottom, with *dihueñes*. You can't even see the branch they're attached to, there are so many. And each one is the size of an apple."

"No!"

"Liar!"

"It's true. As big as apples, and *riquísimas*, too. And *viejos*, let me tell you: I ate those. And I ate, and ate and ate. And since they're so spongy and light, you never feel like you're getting full."

Illanes finishes his story and none of the men have asked him to stop talking about food. "When you're hungry," he tells them, remembering his soldier days on the pampa, "everything tastes good."

•

Ever since his epiphany about God and the need to be strong for his miner brothers, Omar Reygadas has tried to be upbeat around them. God is with us, he says again and again. But the days of hunger, the rising and falling of emotions as he listens for the drills, begin to sap his strength. He is fifty-six years old, a number that hovers over him as he thinks about the pains that are spreading across his body. First it feels as if someone were squeezing his chest, then a burning sensation spreads in his arm, and finally he loses much of the ability to move the arm. He believes he's having a heart attack, and he begins to imagine his own death, and visualizes the thirty-two remaining men being left with his body and how quickly his corpse will rot in this heat. The fear of death grows as he lies on the ground outside the Refuge, the heavy air around him transformed into invisible, suffocating hands. Suddenly, he feels the air moving. It's cooler. There's a fresh breeze blowing over him. He sits up, takes out a cigarette lighter, and watches the flame bend, pointing upward on the sloping Ramp. Air is coming up from somewhere farther down in the mine. Maybe they're injecting air into the mountain. Or maybe one of the drills broke through farther down. Omar announces his discovery to the other men, and a short while later he's on his feet and walking downhill, with a few others joining him in an expedition into the deeper reaches of the mine, to see if they can discover where the air is coming from. The idea that they might find a shaft drilled from the surface and make contact with the outside world drives Omar and the others down past several curves and switchbacks. They reach Level 80 and then Level 70, and the flame is still blowing upward. Finally, they enter the corridor called Level 60 South and here the lighter flame blows straight up and flickers and dies: There isn't enough oxygen to keep the flame burning. At Level 60 North the same thing happens. They go farther down into the mine to Level 40 and there the flame moves back and forth and bends back on itself—the air is moving there, it's fresher again, but then it goes out. They inspect many dark and abandoned corridors but they never do find the opening down below where fresh air is entering the mine. But in all that walking and searching, something else happens to Omar: The tightness in his chest lifts. Thanks to that light breeze, "I started to breathe well again. And when I had to walk back up to the Refuge, the breeze stayed with me all the way back."

Near the Refuge he sees José Henríquez, the Pastor, and tells him what he's seen, how the breeze keeps coming from below.

"Where could it be coming from?" Henríquez says. "The caverns are all blocked up. There is no drill that's broken through."

"It's the thirty-fourth miner, my little compadre," Reygadas says. "He hasn't abandoned us." The thirty-fourth miner is the soul of every miner who's ever toiled, the spirit of the God that protects them.

The cooler air returns every day, at six o'clock in the early evening. "That little breeze [*vientecito*] would come and it would leave us calmer." Omar decides that if he gets out, he's going to tell the world about it one day. "This can't just be forgotten here." All his years in mining offer no explanation that he can think of other than that it's God blowing into the mine. And even if he really hasn't seen a miracle, but rather the product of another shift of rock, it doesn't matter. Because Omar believes that in the bending flame he has seen something divine, again, the breath of God keeping him alive, feeding oxygen to his lungs. He relaxes, takes easier breaths, feels better.

The drilling grinds on, and it then stops, often for hours at a time, leaving a cruel silence that's filled, as their ears adjust to it, with the sound of their breathing and their coughs. When the drill goes quiet the self-described athlete Edison Peña thinks: *This is insane!* A man next to him says: "What are those guys doing up there?" Edison asks himself the same question. He is a sensitive, articulate man who was already deeply in tune with the whole absurd cycle of human existence even before the guillotine fell over the Ramp and trapped him inside. He's had prior bouts of suicidal depression, and going down into the bowels of the mountain always felt like a kind of ending to him. "Death was always there in the mine. I knew it, everyone did. You'd try to tell people outside, and they wouldn't believe you. They'd look at you like you were talking about science fiction." For Edison to enter the mine on a regular day was to face the existential truth most men grasp only near the end of their lives: We will all die. Death is waiting for us all the time. Perhaps *this* is his time, and his waiting will finally end: He thinks this, especially, when the drills stop and the silence inside the caverns of the mine goes on for two hours, and then three. *There is no drill coming for us now. They gave up!* Four hours. Five. With his relatively alert and lucid thirty-four-year-old mind he is seeing firsthand what a true kick in the ass it is to be a human being, because he can see he's trapped inside a kind of metaphor about the cycles of life and death, halfway on that metaphorical journey from the sunshine of being fully alive to the permanent blindness and deafness of death. "I felt an emptiness. A vacuum in my body," he later says. Some of his fellow miners try to fill the silence by doing things like honking horns to let people know they're still alive down there in the mine. Edison hears the noises they make and thinks: *How innocent these people are, how naïve. We're seven hundred meters underground! No one can hear us! No one!* Perhaps more than any other miner, Edison feels fate descending upon him, like some angry creature residing in his rumbling stomach, pulling the life out of him from the inside. Eight hours. Nine. There's no drill. No one is coming for them. Edison tries to fight the emptiness he feels growing inside him, to shake it away, and he starts tossing and turning on the floor of the Refuge, his eyes wildly out of focus. To his fellow miners, he seems to be losing his mind.

The truth is this: Edison was already a bit of a lunatic before he came into the mine. And his was not the loquacious, extroverted madness of Mario Sepúlveda, but a darker, lonelier, more morose introspection. More than once during the routine workdays at the mine another worker has pronounced Edison "crazy" for his tendency to violate certain safety rules: for example, the rule that says a man should never walk anywhere alone in the mine. It's a self-destructive, rash thing to do underground, to go off where you might accidentally step into an abyss, or have a rock fall on you, and not have anyone nearby to hear your muted cry for help. His I-don't-give-a-shit attitude and the possessed look in his eyes earned him the nickname "Rambo" from Alex Vega. Edison walked alone in the mine all the time, daydreaming in those fatal corridors, where he once found a massive, murderous slab of fallen rock at a spot he often walked past.

As he waits for the sound of drilling to begin again, Edison is living in a space of physical and emotional desolation. The thunder of falling rock, the textures of the colorless walls, with their millions of serrated edges, and the increasingly fetid air all suggest that he and his fellow miners have been sent here for punishment. *How can God be doing this to us?* Edison thinks. *Why me? Why us? What did I do?* There is judgment, too, in the simple absence of light. "The darkness around us was really killing us," he later says. Edison the electrician has helped Illanes bring a battery and some light to the Refuge and the space next to it. But then there is a moment when the battery fails, and suddenly everything around Edison disappears in complete darkness. "That's when you really feel you're in hell. That's where hell is, in the darkness." Aboveground, Edison was in one of those tempestuous relationships that is colloquially called "hell," in which objects fly across the room, where the love and hate two people have for each other causes them to treat each other poorly. But now he's in a real hell, as he can see when the weak light comes back on again. It's like he's in the catacombs of the purgatory described by a certain devout Italian poet at the end of the Dark Ages. He sees bodies of men sleeping, or not sleeping, fitfully, stretched out on pieces of cardboard, on tarpaulin, their faces painted black with soot and sweat, in the Refuge and just outside, in rows of that tunnel, that defile of stone that leads down toward the hot center of the Earth. "Visually, it looked like my time had come."

Or maybe not. Because after twelve hours of silence here comes the drill again. *Rat-a-tat-tat, grind-grind. Rat-a-tat-tat, grind-grind.* The sound of other men working to reach him provides some comfort, a muted joy for an hour or two, or three. And then it stops again. "The silence just destroyed us. Because you would feel abandoned, alone. Without a positive sign, your faith collapses. Because faith isn't totally blind. We're vulnerable, we're really a small thing. I knew what it was like to feel alone and helpless, to feel there was no way out. Because your faith empties out second by second, it doesn't get stronger as the days go by. People say that, but it's a total lie. You'll find a lot of my companions who make those stupid statements about feeling stronger. I don't know. I hear them and I want to kill them."

Edison wants to live, and to live he chooses to do as little as possible. Some will criticize him and the other men in the Refuge for not leaving that room, with its cheap white tile floor and steel door. But to Edison it's what makes the most sense, really, when you're not eating. To simply wait and rest. "I conserved my energy. I'd go walk sometimes. But then I started to see that my legs wouldn't respond when I had to go to the bathroom. I was starting to feel really fatigued. I had enough intelligence, innate intelligence for survival, to not do anything to kill myself. A lot of us were like that."

Edison is also in a room surrounded by people with hurt and rage, a long lament with many voices. "They'd say, 'If I get out, I'm going to do this, that, and the other,'" Edison says. "They'd say, 'I wish I'd been a better father.' You'd ask someone, 'How many kids do you have?' and his eyes would fill with tears. And you'd look at the man next to you, and you'd realize that he was even more destroyed than you were. And that's the great truth: that in the mine there were no heroes."

They are not heroes, but ordinary men who are afraid, silencing their rumbling stomachs with large quantities of dirty water, and waiting until noon, when they all gather to eat. But first, just before the meal, the tall, balding José Henríquez begins the session with a prayer, and then a few words that serve as a kind of sermon. Sometimes he tells Bible parables from memory. There is, most appropriately, the story of Jonah, who was swallowed by a whale. Jonah was sent on a mission by God, to speak in a certain village, but instead Jonah got on a ship and went in the opposite direction. "Jonah was a guy with a bad temper," Henríquez says, "so God put the squeeze on him." The Lord sent a powerful storm to toss that ship about, and when Jonah's shipmates realized he was the source of God's wrath, they tossed him overboard, where he was swallowed by a great whale. "Disobedience is never good," Henríquez says. Jonah was in the belly of hell and in the "depths," Henríquez says, speaking a word that he remembers from a Bible passage. *Profundidad* is the word in Spanish, and hearing it spoken by a man of God inside the depths of the mine leaves a powerful impression on the mind of the diary writer Víctor Segovia, who will scribble the word down a few hours later.

"I went down to the bottoms of the mountains," the Bible passage reads. "The earth with her bars was about me for ever." Jonah submits himself to the Lord, he says God has brought him up from a life of "corruption," and promises that he will sacrifice unto God "with the voice of thanksgiving." The Lord then commands the whale to spit Jonah out.

Here, in this horrific place, trapped inside stone walls, the message is more powerful than it will be when spoken in any church: It's as if they are living inside a Bible parable, Yonni Barrios thinks.

They have survived two weeks without a true meal, with no certain prospects that they'll ever eat again, and everything that's happening to them seems to have some deeper message. Víctor Segovia never went to church much, but now he's sort of going to church every day, because with each prayer session the sense grows that the union of those thirty-three men is a holy event. Before this accident befell him, Víctor writes in his diary, he'd thought of church as a place where sinners went to seek forgiveness. But Henríquez speaks to him now of a message of hope and love. The Pastor is, by now, a man physically transformed, too: He's shed his shirt against the relentless heat and humidity, cut off his pants to shorts, and walks around in ripped-up boots that look like sandals. Speaking of God with his bare chest and its patches of hair covered in sweat, and with his bald pate and its matted fringe of hair, Henríquez is beginning to look like a crazed mystic who lives in some desert cave, an effect heightened by the fact that when he speaks, he seems utterly convinced of what he's saying. Christ loves you in spirit, the Pastor says, and Víctor later records the Pastor's words in his diary: "Look for him and you will see that he loves you, and you will find peace." For Víctor, this is a revelation. "I see now that people who are thankful go to church, too, and that the people who go there have been touched by the grace of God," he writes.

In another sermon, Henríquez tells the story of Jesus taking five loaves of bread and two fish and multiplying them to feed five thousand people. He then leads them in a prayer that the Lord will find a way to take their small supply of food and make it last longer, because very soon they're going to run out.

"The Pastor would pray that the food be multiplied," Mario Sepúlveda later says. "Afterward, I'd see one of *los niños* walk over to the cabinet and try to peek inside, to see if there really might be more food there."

Instead, each opening of the cabinet reveals less food. The men begin to scavenge around to see what they might find to eat. Yonni Barrios, the man who failed to protect the food against the hungry men on the day of the collapse, sees one pick up a discarded can of tuna and take his finger to it, wiping the inside and licking his finger again and again: Yonni never thought he'd see a well-paid man like him reduced to such a state. Other men begin to go through the trash cans, and when they find orange peels they clean them well and eat them. Yonni himself devours the brownish remains of a pear. "That was good to eat. The hunger was terrible." Víctor has also eaten half-chewed fruit he found in the trash, and on Wednesday, August 11, he writes about it in his diary, remembering how he used to see the poorest of the poor in Copiapó sifting through the garbage. "We ignore it. People think that it will never happen to them and now look at me, eating peels, trash, and anything that is edible." Carlos Mamani, the Bolivian immigrant, scans the ground to see if there are any bugs or worms crawling around: He would grab one if he saw it and eat it. But, just as there are no butterflies in the mine, there are no beetles or caterpillars either. "I didn't see a spider, or even a termite, nothing."

The men are feeling weaker now, it's getting harder to walk up and down the 10 percent grade of the Ramp, and the sense of physical degradation grows as the space around the Refuge fills with the water flowing from the drills that are trying to reach them. The water is turning the ground to mud, and the mud is swallowing their boots when they walk on it, and the vehicles slip and slide when they try to drive on it. Several of the men take one of the front loaders and try to build a kind of levee against the water and mud but it's quickly eroded away. Mario Sepúlveda goes walking through the mud, shirtless and soot-covered, confusion and worry on his face. He stops talking and the men stare at the lonely spectacle of him as he walks off: There is a thickening layer of black hair covering his cheeks to match the kiwi hair on his scalp. He walks up to the mechanics' camp on Level 190 and tells the men there how he's feeling, how disgraceful it is that he's going to die down here, and the men try to cheer him up. Later, back down at the Refuge, he manages to fall into a fitful sleep. Víctor Segovia sees him dreaming, speaking in his sleep, saying his son's name: "Francisco." It's a painful thing to witness: a grown, middle-aged man who longs to be with his son so much, he speaks to him in his sleep. Then Mario wakes up, looking sullen and crushed, the man of so many words suddenly unable to speak any at all.

•

The other men note that Carlos Mamani is especially quiet, having set himself apart in a corner of the Refuge. Days pass without him saying much. To the twenty or so men sleeping near him day after day, the long silences of this very young man with the indigenous face are disturbing and morbid. Carlos is simply afraid and confused. He got trapped on his first day working underground at this mine, and all these men seem to know one another, or be related to one another. They frighten him because they're constantly arguing over whether they're going to be saved and who's to blame if they aren't. "I didn't know who I could trust."

Now Mario Sepúlveda, the same man who's been wandering the mine in a funk, snaps to attention, and looks directly and very intently at Carlos Mamani. With all the others near the Refuge listening, he stands up and addresses the *boliviano*. "Down here with us, you're as Chilean as the rest of us," Mario says loudly. A lot of working men in Chile resent Bolivians in the same way working people in other countries resent outsiders, and everyone knows that being a Bolivian in Chile isn't easy. "You're friends and brothers with all of us," Mario says. The speech ends with all the men around them breaking into applause, and some wiping away tears, because it's true: They're all dying together, and no human being, not even the Bolivian among them, deserves this fate. Carlos has been watching the men spend hours playing dominoes, and now they invite him to join in, and since he's never played dominoes before, they teach him the rules. It's a simple game—twenty-eight tiles, match number to number, etc., etc.—and Carlos learns quickly. He sees how these endless rounds of dominoes can make the nights shorter, the darkness less dark. After a few games, he wins one. And then another. Pretty soon, he's beating everyone.

"He won? Again? Who taught this Bolivian guy to play?"

In Chile, among men, when you really are brothers with someone you mock them. This is called *echándole la talla*, which can be translated roughly as "taking his measure." Being able to mock someone without causing a fight is a valued skill, and among the men Víctor Zamora is best at it: That's one of the reasons hardly anyone is really all that angry at him anymore, even though he did lead the raid on the food supplies and thanks to him they're hungrier than they need to be. At any given moment, Víctor has got half the men in and near the Refuge laughing at the other half. *Look at that Mario Gómez, with his block of wood, listening to the walls,* he might say. *Is the drill nearby, Mario? Which way is it coming from?* And then Víctor will stand up and point the way Mario Gómez does, like some Labrador retriever. Sometimes, if Gómez isn't looking, Zamora will point with his hand showing just three fingers—the miners know it's mean, of course, to mock a man with a maimed hand, but in the context of this cave it's very funny. *From over here! No, from over there! It's close!* Zamora's jokes at the expense of Gómez are so funny they keep the men repeating them and laughing for days afterward.

Eventually, to bring Carlos Mamani into the fold, Mario Sepúlveda directs a bit of ribbing at him, too. Like the jokes the men tell about the others, it draws on the quality that sets him apart.

"Mamani, you better hope they come for us. Because if they don't, since you're Bolivian, you'll be the first one we're going to eat."

Mamani isn't especially bothered by the joke—does anyone take any of these Chileans seriously? "I never thought they were going to eat me," he later says. But when Raúl Bustos hears this joke he thinks: *Now that crazy Mario has gone too far.* So do a few other men. What kind of lunatic jokes about eating someone to men who haven't eaten a real meal for ten days now? They really are starving to death, some of the men think, and it's not outside the realm of possibility that they might have to eat the first guy that dies. "I know Mamani didn't sleep well that night," Florencio Avalos says. Raúl is also troubled by this macabre humor: He's not entirely sure these men will all hold together if they truly begin to starve to death. After the tsunami in Talcahuano, people very quickly surrendered to their baser instincts. Mario Sepúlveda is a man of swinging moods, and Raúl senses now what Mario's closest friends and family know about him: He's a man who is not entirely in control of his emotions. He'll say he loves you one moment, and threaten you the next, and it seems likely that he will do anything to survive.

Even as they all grow weaker, Mario is picking fights with people. He argues with Omar Reygadas over the drilling and the drillers. The older Omar has worked with drill crews before, and each time the drilling stops, or when it seems to be going off course, he shares a bit of this experience. In the middle of one of those long silences when no drilling can be heard, Omar tells the men around him not to worry, reminding them, again, that he's worked with drill crews and knows a bit about their routines. "They're not giving up," he says. "It's just that they have to reinforce the bars . . ." By now the men are experiencing new, unmistakable symptoms of malnutrition and starvation. Walking to the spot where they go to the bathroom takes effort, and when they reach that spot, it's often the site of a squatting torture. Their bodies want to push something out, but it takes too much agonizing effort, and what finally does drop to the ground is strange looking. Their feces are compact, oval-shaped pellets, as hard as stones, and to the men who've grown up on farms and lived in the country, they look like goat or llama droppings.

Mario Sepúlveda is as constipated, exhausted, and freaked-out as the rest of them, and finally he decides he's had enough of this white-haired bullshitter. "You're always saying the same thing!" he barks at Omar. "You're lying. You don't know anything. You're an idiot!"

"You can't talk to me that way."

"Shut up already!"

Omar protests the attack on his honor by standing up and taking a few threatening steps toward Mario, unconcerned that the man with the heart of a dog is taller, stronger, and younger than he is. "Let's go and settle this . . . down by the water."

As several men watch, the two men walk away from the sleeping area near the Refuge and begin to walk down the Ramp. They're headed toward a side passage where there's a pool of water, built up by the trickle flowing into the mine since the rescue drills began working. Mario walks and thinks of the violence he'll inflict on this annoying man, how he'll finally let loose this anger that's been welling inside him. But the pool of water is more than a hundred meters away, and in the minute or two it takes to walk this distance, his anger lifts. The older man seems really determined to fight, he's not going to back down, and looking at him, Mario realizes Omar is as desperate and hungry as Mario is, and how absurd it is to be fighting in this place when they're so close to death already.

"I looked at this guy, who was older than me, and I thought: *If I, the young ram, beat up this old goat, I'll have a lot of explaining to do. And if this old goat beats up the young ram, then I'll have even more explaining to do.*" As they face each other by the pool of water, each man's helmet illuminating the other man's face with a beam of light, Mario breaks into a mad grin. He shares his observations about young rams and old goats, and apologizes to Omar and wraps the older man in a sweaty and heartfelt embrace. They're starving and they're going crazy, but they're still brothers. "I'm sorry, *viejo. Perdóname.*" Forgive me. Omar looks relieved, exhausted. They walk back to the Refuge. As they approach, the other miners stand or sit up straight at attention, expecting to see two men who've pummeled each other. Instead they see two bare-chested, soot-covered, hungry miners laughing and joking like fast friends.

Juan Illanes has installed lights near Level 105 and Level 90, but the sense that the men are surrounded by forbidding darkness grows as the days pass and many of their lamps dim and go out altogether. The prospect of being surrounded by complete darkness causes Alex Vega to remember a miners' legend: Men left in the dark for too long will eventually go blind. And Jorge Galleguillos remembers a few times in his mining career when his lamp stopped working and he found himself in total darkness: You can get disoriented quickly, and it can be frightening to be lost and helpless, reaching out with your hands to try to find the cavern wall you remember is nearby. Finally, Illanes discovers that he can charge some of the batteries of their lamps using the generators of the vehicles trapped below, and the dark isn't quite as forbidding after he returns the light-giving devices to them.

Eventually, the doers among the men decide they just can't sit and wait for the drills to reach them. The rescuers will eventually give up without a sign of life from below. So the trapped men renew their efforts to send a message to the top. They have dynamite and fuses, but not blasting caps, since no blasting had been planned for the day they were trapped. But Yonni Barrios and Juan Illanes come up with a plan to make blasting caps by extracting the black powder from the fuses and using the foil inside the discarded milk cartons to collect and concentrate it so that it will ignite and set off the nitrate-based explosive they use for everyday mining work. They walk up as high as they can in the mine and set off Yonni's homemade detonator, waiting for 8:00 a.m., when the drilling stops every day for what is clearly a change in shifts up there on the surface. When the silence comes, Yonni lights a fuse leading to his improvised blasting cap. It works, the dynamite explodes, and the explosion is a powerful one—but no one on the surface hears a thing.

We're seven hundred meters underground, Juan thinks. *How could they hear anything?*

When the drilling starts again, the sounds get closer and closer, the vibrations and the pounding palpably close in the rock. The miners say things like "This one belongs to me" and "This is the one that's going to burst through." They go up and down the levels of the Ramp and the side passageways looking for where the drill might come out. Then the drilling gets farther away. It stops.

On August 15, their eleventh day underground, Víctor Segovia notes in his diary the many signs that he and others are losing hope. "It's 10:25 and the drilling has stopped once again. Again they sound really far. I really don't know what's going on up there. Why so many delays? . . . Alex Vega yelled at Claudio Yáñez, who sleeps all day and doesn't co-operate with anything . . ." There's still work to do: primarily, gathering water from the tanks at the upper levels. The next day Víctor writes: "Hardly anyone here talks anymore." On August 17, he sees miners gathered in a small group, murmuring. "They are starting to give up," he writes. "I don't think God would have saved us from the collapse just to let us die of starvation . . . The skin now hugs the bones of our faces and our ribs all show and when we walk our legs tremble."

The drilling stops for several hours, and the men wander the mine searching for any sound, and then it starts again. The thing inside the rock pounds and grinds for a day, and suddenly it's quite close, and the men start to talk about the preparations they've discussed before. They find a can of red spray paint, used in routine mine operations to paint a red square or circle on the mine wall indicating the path to the surface. If the drill bit breaks through, they'll paint it, and when the drill bit operator pulls it up, he'll find that unmistakable proof that there are living men down below. José Ojeda once worked at El Teniente mine, the largest underground copper mine in the world, and in the safety training he received there he was taught to include three pieces of basic information in any message left for potential rescuers: the number of trapped men, their location, and their condition. With a red marker and a piece of graph paper, he now prepares such a message, boiling it down to seven words. Richard Villarroel, the expectant father, hunts about his tools for the hardest piece of metal he can find, and comes up with a big wrench. If and when the drill bit breaks through, his job will be to pound on its steel casing with this tool, making a sound loud enough to travel up the two thousand feet or so of metal leading to the top, where some rescuer might have his ear to the pipe, listening for a sign of life from below.

After a day it becomes clear the drill they're hearing is actually beneath their feet, and they try to follow its sound, walking and driving deeper into the mine, listening in one of the twisting passageways farther down, going lower and lower until it fades away. On August 19 the diary keeper, Segovia, writes: "We are getting desperate. One of the drills just went by the walls of the Refuge but it didn't break through." The following day he notes, "Perri's spirits are very low." The only sustenance for the men that day is water, because the food is running out and there's only enough now for a cookie every forty-eight hours. "The drill does NOT break through," Segovia writes the next day. "I'm beginning to wonder if there's a black hand up above that doesn't want us to get out."

The trapped men have now heard at least eight different drills approach them, only to stop, or fade away in the distance. Several of the miners follow the last drill down several levels, and listen, disbelieving, as it passes below the lowest spot in the mine, Level 40. "That was horrible. That was like a second death," one of the miners says. The idea that they've been doomed, again, by the mine's owners becomes a real possibility: The San Esteban Mining Company's blueprints are so unreliable that the driller-rescuers up on the surface will never find them. "The mine's blueprints are shit," they shout. *La planificación de la mina es una hueva.* The thirty-three men now sit in the dark, wondering if they'll die suffering this final assault on their dignity: trapped here, starving, with other mining men working to reach them, their efforts betrayed by a company too cheap to even know, with certainty, where its own tunnels are.

9

CAVERN OF DREAMS

Laurence Golborne, the minister of mining, is desperate, and he's getting all sorts of crazy advice. His drills are missing their target, or the drill bits are breaking before they reach the level at which the miners are trapped. More than a dozen such holes have been drilled, each a failure. On August 19, with the men having been trapped for exactly two weeks, one of the drills passes 500 meters. It's headed for one of two open galleries in its path, and Golborne and André Sougarret and the others are optimistic they will hit something. The families have been informed that the drill is close, and a series of hopeful all-night vigils begins at Camp Esperanza—but the drill just keeps on going, never finding an open passageway, and eventually it will reach 700 meters without striking anything. "The driller was so emotionally invested he couldn't stop, even though we knew he'd gone too far already," one official says.

Golborne tells the press that he isn't sure what's gone wrong, though he suggests the mine owners' diagrams could be inaccurate. Sougarret, the lead rescuer, says the same thing to the newspaper *La Tercera*: "With bad information, it's hard to make decisions." Another, unnamed government source tells *La Tercera* that it's possible the entire mine has collapsed, one of many pessimistic statements that filter down to the families. "That night, there was a revolution among the families," Golborne says. You don't know what you're doing, they say. Codelco has no idea what it's doing. We know! Listen to us! A union of small miners, *perquineros*, announces that they will enter the mine, "on our bellies," *de guatitas*, if only the government will open the sealed-off mine entrance.

Finally, at the behest of some desperate relatives, the minister of mining agrees to talk to a few "wise" people who the relatives believe might be able to help. One is a psychic, and Golborne meets with her on a freezing cold night. "I see seventeen bodies," she says. "I see one whose legs are smashed. He's screaming." Golborne decides it's best not to pass on her "findings" to the families, who have also insisted he talk to a "treasure seeker," a man who uses a kind of secret divining-rod "technology" to study the surface of the mountain where the men are buried.

"What kind of technology is this?" Golborne asks.

"Well, it's very complicated," the treasure seeker says.

"I'm an engineer. Explain it to me. Is it based on sound waves, heat, voltage differences?"

The treasure seeker says it really is too complicated and declines to explain further, but Golborne grants him access to the mountain anyway, to please the families more than anything. The treasure seeker spreads long rugs across the surface of the mountain, he takes some measurements with a device Golborne has never seen before. When he's done, he announces to Golborne in a haughty tone that his drilling teams are looking for the miners in the wrong place. The treasure seeker says Golborne, Sougarret, and the rest of their team are a bunch of idiots who are going to let those thirty-three men die if they don't listen to him and pay attention to what his instruments are saying: You won't find them unless you drill in a totally different area.

Golborne ignores the treasure seeker's advice and walks down the mountain to sit with a woman who earns a living selling pastries by the beach, a woman who has earned the trust of the families and whose trust he must earn, too. He has to make her believe that he's trying everything that can be done, that he and the government are using every resource and all the knowledge at their disposal to find the trapped men. He speaks to María Segovia, the sister of Darío and the "mayor" of the family camp. María has heard the news of the drill passing 530, 550, 600 meters, each number another blow. "We're running out of time," she tells him. She repeats the phrase. *Se nos está acabando el tiempo.* There are more drills working, he tells her, though his face betrays both exhaustion and concern. We have not given up.

María Segovia will remember that moment with the minister as her lowest. "You have to fight and fight, but at the same time, you feel this sadness, this worry, this sense of powerlessness," she later says. She's bundled against the cold, listening to the minister in his official red jacket. GOBIERNO DE CHILE, it says in white letters. The minister often comes down to her camp, and sits with her and her family and drinks maté tea with her, and in this way he's won more of her trust. The minister acts oddly humbled in her presence, and he says another drill is just two days or so away from its target. María is fighting her natural skepticism for the privileged and everything they say, and she's trying to believe him.

The average human brain requires about 120 grams of glucose each day to survive. The thirty-three trapped men are ingesting, on average, less than one twentieth that amount. During a man's first twenty-four hours without steady food, his body produces glucose from the glycogen stored in his liver. After two or three days the body begins to burn the fat stored in his chest and abdomen, and around his kidneys and many other places. His central nervous system cannot survive on such fats, however. Instead, his brain is fed the acids, or ketone bodies, produced by his liver as it processes his body fat. When his body's fat reserves are exhausted, the protein in his body—muscle, primarily—becomes his brain's chief source of energy. The body's protein is gradually broken down into amino acids that the liver can convert to glucose. In effect, a man's brain begins to eat his muscles to survive: This is the moment when starvation begins. After two weeks the smaller and thinner of the thirty-three men trapped in the San José have lost enough muscle mass that their colleagues begin to notice.

Alex Vega's clavicle is starting to push out against his skin. "Hey, Bicycle Chassis, look at you!" Omar Reygadas says to Alex, the man who came to the mine so that he could add some rooms to his house. Then Omar thinks that, no, a bicycle chassis is too heavy and big a metaphor to describe the way the shirtless, thinning Alex looks. He looks like *charqui*, a Chilean idiom roughly equivalent to "jerky." "*Charqui* is what animal meat looks like when you dry it up." *Charqui de mariposa*, he calls Alex. Butterfly jerky. "You can imagine what butterfly *charqui* would be like. That's basically just dust."

Alex takes this with the humor and endearment with which it's intended. Omar isn't looking so great either, after all. None of them are. Their metabolisms are slowing down, even the most energetic among them are sleeping longer than normal, and there is a haze starting to drift over their thoughts. Several are beginning to experience one strange, unexpected side effect of prolonged hunger that's been noted again and again by people who fast for a week or more: When they sleep, their dreams and nightmares are unusually long, vivid, and lucid. Their dreams seem more like real life, an effect many devoted fasters attribute to the purification of the body and brain during fasting. Deprived of sustenance, their brains take the men to places of memory and desire, mind dramas crafted from the material of their personal histories, with a cast drawn from their families and loved ones.

Carlos Mamani, newly crowned domino champion of the ongoing game in the Refuge, finds his subconscious setting out on a series of journeys. "I would sleep not to feel the hunger," he says later. "Then I'd dream and in my dreams I'd go to see my siblings. I'd wake up a bit, fall asleep for a long time, and I'd see another of my brothers." His ten brothers and sisters have been dispersed across Bolivia: from his home village of Chojlla in the province of Gualberto Villarroel, to the big cities of La Paz and Cochabamba. The Mamanis were all orphans, and raised one another. "The only one I didn't see was my older sister, the one who made me study after my parents died. I went to their houses. One right after the other. I went to see my aunts and my cousins, too." In his dreams he's walking on the Altiplano, down those unpaved roads, past corrals for llamas and goats, and into small living rooms crowded with furniture in big cities, or back in his village, where the glacier-covered peak of Illimani, the "Golden Eagle," is visible in the distance. He grew up on that plain, where the people grow potatoes, oats, and the seeds *cañahua* and quinoa. "I grew up in the *campo*, in the provinces," he says. "In the countryside they say that when someone is about to die, they walk at night. In my dreams I was walking." When he wakes up, the implications of the dreams sadden him: He isn't prepared to die so young. He remembers his days as a schoolboy in Chojlla, an orphan going to school at the insistence of his oldest sister. It was a long walk across the Altiplano back home, and he might get home at 8:00 or 9:00 p.m. Some nights, on these boyhood walks, he'd see the silhouette of a person in the blink of an eye, and then it'd disappear. It was the spirits of those about to die, he thought, and now he's one of those people, his spirit wandering when he dreams. He sees one sibling after another in a chain of dreams, but he never sees his older sister, the one who helped raise him when their mother died when Carlos was four. He remembers the day of his mother's funeral, what seemed like a very long party with food and kids running around, but the memory of growing up with his older sister getting him ready to go to school is much stronger. He has not yet and will not dream of his sister. "I figured if I saw my sister in one of my dreams, that would mean I really was about to die." Instead, Carlos has another, hopeful dream. He's standing in a huge metal bucket—like the one in a mine he used to work in—being carried up to the surface, like a man riding an elevator to the top, to sunshine and safety.

Eduardo Hurtado and the Terraservice team, fresh off the failure of their previous hole, are now drilling hole number 10B. They had begun this, their third attempt to reach the men, on Tuesday, August 17, before dawn. Every hundred meters they stop drilling and the topographer, Sandra Jara, lowers a gyroscope into the shaft and measures their progress. Jara and Hurtado and the drill operators consult with one another, and as the shaft gets deeper they combine their collective knowledge of the science of topography and the craft of drilling to reach a critical decision: They will drill very slowly, sacrificing speed for accuracy, at just 6 rotations per minute, less than half the usual 12 or 15. Nelson Flores, one of two men operating the drill for twelve-hour shifts, understands the need to do this, though it goes against his instincts. "You get bored, you want to make it go faster, to get the job done," he says. When his twelve hours are done and he joins the other men passing through the gate, they get a round of applause from the families waiting there.

On one of those nights when the skipped meals and the distant drills eat away at his soul, Edison Peña moans for several minutes about his imminent death as he tries to sleep. "I'm dying, I'm dying," Edison says. Mario Sepúlveda is trying to sleep next to him, and is at his wits' end listening to this. *Enough, Edison,* he thinks. Finally, the trickster in Mario kicks in. He tosses his head back and forth in imitation of the dying Edison, mouth open to make a choking, gurgling sound, as if he were beginning to suffer a final, starvation-induced seizure. He launches into a movie-like death speech—Mario loves movies, especially anything with Mel Gibson in it. "This is the end, Edison," he moans weakly. "I'm dying. I'm going. Tell . . . my . . . wife . . . that . . ."

When the actor Mario closes his eyes and goes silent, Edison sits up, leans over Mario's chest, and starts to shake him desperately.

"No, Perri, no!" Edison cries out. "No! Don't die!"

Mario opens his eyes and breaks into a wicked smile, a peal of laughter, and then a few choice vulgar Chilean idioms about foolish men. Mario thinks that his fake death scene is one of the funniest things he's ever done. Edison starts to act as if he were in on the joke, and a bit later, in fact, he and Mario repeat their death sketch with Edison throwing in the line: "Perri, tell me where you hid the money! Where's the money?" Another miner who witnesses these death scenes says: "At first it started out as a joke, but after a while it started to seem real." You'd think that men who are as close to death as they've ever been wouldn't be able to joke about it, but Mario and Edison are different. Edison explains it this way: "I think that sometimes the only thing that can make you laugh is accepting the idea that there's no way out." For a short while after their first, absurd little show, the suffocating veil of imminent death inside the Refuge is lifted by the memory of Mario's laughing at Edison's expense. Who else would do such a thing? Who would mock a man's death lament? The same man who can call them all to prayer, or who can tell a group of starving men that he's going to eat someone.

Up on Level 105, where the mechanics sleep, the level-headed, good-natured Juan Illanes, the man with the deep radio voice, is keeping up the spirits of the men around him with his steady storytelling and his expositions on a series of topics. On an ordinary day, he can be kind of a bore, but to the trapped men his endless talking is a welcome distraction. He knows a few men are worried about their loved ones and whether their wives and children will or won't survive in the absence of their incomes, should it come to pass that no drill ever reaches them. Raúl Bustos, who has two young children, and Richard Villarroel, whose wife is expecting his first child, are especially distraught, so Illanes begins to expound on Chilean labor law.

"Let's say we don't get out—hypothetically speaking, of course," Illanes says. "The labor laws with respect to social security and accidents are very specific. There's an insurance plan. I'm not sure exactly how much it is. But it's around two thousand UFs. Or maybe three thousand."

"Really?"

"That much?"

The men forget, for a moment, about their difficulties and start doing the math in their heads. A UF is the "Unidad de Fomento," an exchange rate adjusted to inflation that's used by the Chilean government for certain financial transactions; at that moment, a UF is equivalent to a little more than 20,000 Chilean pesos, or about $40. So Illanes is telling the men their families will get anywhere from $80,000 to $120,000, or nearly a decade's worth of wages for your average Chilean working stiff.

"But that's not all," Illanes continues. "Your widow, hypothetically speaking, of course, is entitled to receive your salary if you die in an accident. It's law sixteen thousand, seven hundred, and forty-four." Illanes claims to know the law's exact number (the correct one, as it turns out). *Dieciséis mil setecientos cuarenta y cuatro.* The sound of that number, like the details in his stories about barbecues and exotic South American fungi, makes what he's saying sound all the more real. "According to this law, what you get is calculated based on the average of your pay the last three months. Your wife gets this until she's thirty-five years old. But if your kids stay in school, if they go to college, she'll get it until she's forty-five. And by then, let's face it," he says with a wicked smile, "she'll probably have found some other *viejo* to take of her." How does Illanes know these things? "If you read the law, it says the company has to inform you of all this," he tells the men. "And I've worked so many different mines, I've heard it over and over again so many times I finally memorized it."

He sounds like a lawyer, this Illanes, and not like the mechanic he is, and for a while his confident summary of Chilean labor law leaves the men around him a bit calmer in the knowledge that they really will be able to provide for their families even if they never leave the mine.

When Illanes finally stops talking and is left with his own thoughts, he occupies his mind with small tasks. He imagines he is aboveground, about to resume his normal life, with mundane household chores awaiting him. In the sort of mind game that kept the French prisoner Papillon sane while in solitary confinement on Devil's Island, Illanes transports himself away from the prison of the mine, to his own home, where he left the materials for a table he was building. Now he assembles that table and does other things. "I need to fix that leak in the ceiling. So I'm going to have to fix that gutter. I'm going to have to buy three gutter channels, and two meters of downspouts. How much will that cost me?" He performs the calculations in his mind again and again, imagines each wood screw and fastener and the tools he needs. He climbs up the ladder outside his home with a drill, and then he climbs down again, and when he's finished he does it again. When this is no longer satisfying he tells himself he needs to remember those church hymns from when he was in the choir at fourteen. Illanes sings at his church in Chillán, but he wants to sing an old hymn he hasn't sung in years. He can remember only a few words from the beginning. *"Quiero cantarle una linda canción . . ."* How did the rest go? For three nights, Illanes strains his memory in search of the rest of the lyrics. Slowly, bits and pieces come to him, and this, too, is like building something. By the fourth night he's remembered the whole thing, all four stanzas, all sixteen lines, including the last one, "Only in him did I find happiness," *Sólo en él encontré la felicidad*. With the completed song in his head, Illanes goes off to one of the many passageways in the mine, alone, to a spot where no one will hear him. He sings the hymn out loud, like the teenager he once was, and weeps as he does so, because he realizes now how beautiful it was to be young and to be asked to sing.

On the fourteenth and fifteenth days underground, even the men who've kept the busiest and worked the hardest begin to surrender to exhaustion and hopelessness. For two weeks Florencio Avalos, the second-in-command to Luis Urzúa, has driven up and down the mine, gathering water, looking for passageways out, trying to send messages to the top. He's been going to the Refuge, too, to try to keep up the spirits of his younger brother, Renán, who has spent nearly all his time lying on his makeshift bed there. Get up, Renán, he says, cooperate with something, get out of here, it smells horrible, he says, and sometimes Florencio succeeds in getting his brother on his feet and working for a while. The unspoken truth is that Florencio has been worried that his younger brother might take the miner's traditional way out of a desperate situation: leaping to his death in El Rajo. When a man stands over this abyss and shines a common miner's headlamp into it, he sees nothing but blackness. Killing yourself in the Pit is like jumping into a black hole. A fall of just ten feet or so can kill you in a mine, but in the Pit you can fall one hundred feet. A few of the men will confess to thinking about this kind of death as an escape from the unrelenting aural torture of the mountain, its constant thunderclaps of falling rock.

Finally, it's not Florencio's brother Renán, but Florencio himself who feels utterly defeated. Florencio, the shift's foreman, or *capataz*, is one of the few miners universally admired by his peers—"our *capataz* is young but an extraordinary person," Mario Sepúlveda will say while the men are still trapped underground, "a man who is always overcoming obstacles [*tirando para arriba*] and who has beautiful qualities." But Florencio begins to lose his fight with despair on the night he falls asleep on a bed of big rubber tubes and wakes up to find water rushing over his legs. He rises to his feet and finds himself flailing in mud that moves over his boots in sticky waves. Walking is literally a slog, and even when he tries to drive over the mud the wheels of his pickup truck spin and slide and refuse to climb, adding to the sense that it's useless to try to do anything to rescue himself and the others.

Florencio is walking up with a team that's driving uphill with the bucket truck the miners use to transport water when the futility of his situation finally becomes too much to bear. He decides he cannot and will not take one more step, and he drifts off toward a parked truck without the men ahead of him noticing. He enters the cab, and as the members of the water crew walk away and their lights dim, he's left in the dark because the battery to his own headlamp is dead. *Hasta aquí llego,* he thinks. He's reached the end of his journey. Florencio leans back in the truck's seat, exhausted. The truck's battery was removed to bring light to the Refuge, and Florencio has deliberately placed himself in a state of blindness and helplessness. He feels weaker than he's ever felt before. Let starvation take me away here, he thinks, on this cushioned seat with the windows closed, away from the mud and thunder. Alone in absolute darkness, he surrenders to the idea that he'll fall asleep and never wake up again. He begins to think about his children and imagines them growing up in his absence: César Alexis, "Ale," who is sixteen, the boy he and Mónica had when they were teenagers; and Bayron, who is seven. What will they look like as men, his two sons? How will the passage of time unfold in his absence, how tall will they grow, what will they achieve? Will they start families and homes of their own? It's easier to imagine Ale as a man because he already is one, almost, responsible and studious. Florencio's death in this mountain will help make one thing about his sons' future certain: Neither of them will ever work in a mine.

The other men in the water detail eventually notice Florencio's absence. They look for him in the corridors, in the Refuge, and at the spot where the men go to the bathroom, and they don't find him.

Florencio falls into a deep sleep. When he wakes up, he doesn't feel quite as desperate. Eventually he sees light. He sits up in the pickup and soon the beams of the search party's headlamps are shining on his face.

"Here you are, Florencio."

"We were worried about you."

"We thought you threw yourself in El Rajo."

As the sixteenth day without news arrives, the wives and girlfriends and children of the trapped thirty-three men are also allowing themselves to imagine what the future might look like if the men are never rescued. Elvira Valdivia, the wife of Mario Sepúlveda, has been staying at a hotel in Copiapó, praying with her daughter and son at their bedside every night. After the latest prayer, her son, Francisco, asks her: "Are you sure my father is still alive?" He's twelve years old but he poses the question like a grown-up, like someone who needs an honest answer.

Yes, his mother says. But there is, perhaps, a trace of doubt in her voice, a sense that she's finally losing hope, because now Francisco asks another question.

"What if he isn't?"

Elvira thinks about this a moment and answers: "Son, you have to be prepared for anything because if your father isn't alive it's because

God wanted it that way. Maybe his life only reached this far and we're going to have to learn to live without him. Whether it makes sense or not, that's the way it might be."

"Damn, Mommy, that would hurt," he says. *Pucha, Mami, qué lata sería.* "But what can we do?"

The possibility of Mario's death is a discussion Elvira can't and won't have with her eighteen-year-old daughter, who seems to be holding on by a thread. Scarlette is taking medication to sleep, and asking her mother unanswerable questions. "Does my father have water, does he have light?" To her daughter, Elvira does not express the slightest doubt that Mario is alive. But Francisco wants to know the truth, he's steeling himself for a future that unfolds without the man who is his hero. Francisco is willing to do this, clearly, because his father raised him to be a "warrior" and to deal with painful truths as a man should. Elvira can see now that Francisco has the strength and determination of his father, but also a calmness Mario doesn't have. When Francisco was born he weighed just 1.09 kilos, about 2.4 pounds, and it's one of those everyday miracles of the human species that a boy so fragile and small when his mother first held him can grow up to become a young man possessed of the inner strength to lift up his mother's spirit and help her prepare for a future without the man she loves.

At about this time, deep in the mine, Mario Sepúlveda is still in charge of dispensing the daily meal, such as it is—among other things, it's no longer daily, and it isn't much of a meal. Breakfast, lunch, and dinner are now a single event that comes once every two days. A cookie that can be split in half. At the end of one of these meals, there's dessert. A single slice of peach, about the size of a thumb, has survived an earlier dispensing of the contents of a can among the men. This single peach slice is precious—and it has to be divided thirty-three ways. An act of surgery is required, and Mario performs it slowly, with several men staring. "Excuse me, Perri," one of them says. "But isn't that one piece there bigger than all the others?" When Mario is finished, each miner takes a sliver about the size of a fingernail. Like most of the other men, Mario allows that hint of syrup and fruit to linger on his tongue like a communal wafer, trying to hold it for as long as possible, managing to keep it there for quite a long time—until another miner bumps into him and he accidentally swallows his morsel, and he wants to slug the guy, he's so angry.

But mostly, there's just that one cookie, with its approximately 40 calories and less than 2 grams of fat. It isn't enough to keep them alive, and Víctor Zamora, the man who led the assault on the food cabinet fifteen days earlier, can see this. "It's the most terrible thing," he says. "That's what I'll never forget: to see your *compañeros* dying before your eyes."

By now, the daily prayer sessions and meetings have evolved into ever-longer apology sessions as well. I'm sorry I raised my voice, a miner might say. I'm sorry I didn't do my part with the water yesterday. Today it's Víctor Zamora, with his round face that isn't so round anymore, his curly mop of hair weighed down by sweat and grime, who steps forward.

"I want to say some words to the group," he begins. "I made a mistake. I was one of the people who took the food out of the box. I'm sorry. I regret that I did that." Not all of the men knew about Víctor's role in the disappearance of the food, and some now find out for the first time. "I thought we were only going to be stuck here for a few days," he continues. "I didn't realize the harm I was causing by taking that food. Now I truly regret what I did and I'm sorry for it." Víctor seems very nervous, deeply regretful as he delivers this speech in a soft, tremulous voice, Omar Reygadas later says. "We all realized that he really did feel terrible about what he had done."

After the apology, it's time to eat. This is a day when they'll eat. But then Alex Vega steps forward. "Can I speak?" he asks. El Papi Ricky has become Bicycle Chassis, or Charqui de Mariposa, and he looks smaller and frailer and clearly more in need of a meal than most of them.

Mario Sepúlveda turns to Omar Reygadas and whispers: "This guy is going to ask for more food. What do we do?"

"I'll share a bit of my cookie with him, you another little bit," Omar says. "We'll ask if there's anyone else who wants to help out . . ."

But Alex isn't asking for more food. "This thing is going to go on for a while," he begins. One drill has just missed them and it's possible the next one they hear coming will, too. "There's only a little food left and I think that today we shouldn't eat. Let's not eat. Let's leave it for tomorrow, and that way we'll last a day longer."

Some of the men groan and shake their heads: No, they don't want to skip their meal. Let's eat! I want to eat! But in the end they do skip it. Three days without anything but water. Several are deeply moved by Alex Vega's act of nobility, his willingness to sacrifice, the skinniest man among them putting the group and their collective health beyond his own obvious need.

After listening to the sound of that last drill missing them, several more men begin to write farewell letters, in imitation of Víctor Segovia. Like him, they write in the hope that some rescuer, one day, might find their final message. They're starting to feel weak enough, now, that it seems possible that the next time they fall asleep they might not wake up again, or that they might soon lack the strength to write. Some of the men need help to rise to their feet and walk to go to the bathroom, holding each other up to climb up and down the slope of the Ramp, to the pile of rocks where they bury the llama pellets that come out of their bodies, and to the nearby rank-smelling porta-potty. Someone suggests that they reconnect the hoses that lead up to the water tanks, because in a few days they will be too weak to travel up higher in the mine and fill the barrels of water they've been bringing down to the Refuge. The sense of finality is contagious, and as the writing of goodbye letters spreads, Carlos Mamani watches and listens to the emotional Chileans around him call back and forth to each other: "Are you done? Lend me that pencil. I need some paper." Some of the men weep as they write, and Mamani hears them, and feels bad for them, because for a mining man to cry in front of all of his coworkers like that, he must be truly broken. To Mamani, it seems that the older miners especially are beginning to resign themselves to their fate. "I heard people say later that the older guys were all pillars of strength down there, but that's a lie," Mamani will remember. Jorge Galleguillos has a swollen foot. Mario Gómez's lungs are barely holding up. "The only one of the older guys who was always strong was Omar Reygadas. He'd always say, 'Don't worry, they're coming. They're coming for us.' But most of the old guys started to go crazy." Víctor Segovia has been writing about his mortality for days, and now that most of his companions are doing the same, he finally allows some of the dark thoughts he's been committing to paper to spill out into the open.

"We're all going to die!"

"Be quiet, old man! ¡*Concha de tu madre!*"

Carlos Mamani resists the temptation to begin his own goodbyes to his loved ones. He hasn't yet dreamed of his oldest sister, and until he does, he won't believe it's his time to die. "I didn't want to write my letter . . . If I started to write that letter, it would be because I was in my death throes." Mamani feels weak, but he is not yet in his final agony, *agonizando*. And besides, even if he wanted to write a letter he doesn't have a lamp to write one, because he left it up in the changing room.

Mario Sepúlveda is still strong and alert enough to see how degraded the men in the Refuge are becoming. He picks out the slight Claudio Yáñez as looking especially immobile and pathetic. Yáñez is a small man with angular features. His cheeks have hollowed out, sharpening a haunted look, a faraway gaze. Mario can get the rest of the men to sit up to the sound of his voice, but Claudio just lies there.

"Hey, *concha de tu madre*, stand up! You have to stand up because if you stay tossed there on the floor you're going to die, and we're going to eat you. For being lazy we're going to eat you." When spoken by a man who hasn't eaten in three days, the words *we're going to eat you* carry a meaning they might not otherwise. "So you better stand up, because if you don't, we're going to make you stand up by kicking you." Startled, Claudio tries to climb to his feet, and when he does everyone around him can see how skinny and weak he's become. He rises slowly, with buckling knees and bent legs. "It was like watching when a little horse is trying to walk right after it's born," Omar Reygadas later says. Finally, the "little horse" straightens his legs and takes a step.

The younger men like Claudio are in bad shape, too, and have each lost about thirty pounds. When Alex Vega stands up to go to the bathroom, his vision clouds and then, for a few seconds, he goes blind, the first signs of a common side effect of hunger, caused by vitamin A deficiency. Many of the older and bigger men still have layers of fat around their waists: It's their upper bodies that have caved in the most, giving them a boyish appearance when they walk around shirtless. They can now see that the beefiness in their chests wasn't composed of muscle after all, but just the layers of fat of the overfed. But it's in their faces and expressions that the change is most dramatic. Yonni Barrios has eyes that have retreated into his skull, his once-seductive brown irises fixed in the sad stare of a man suffering something akin to combat fatigue. The veteran miner and onetime whistle-blower Jorge Galleguillos opens his mouth to talk and seems to be chewing his words as he does so. To keep the weakened Jorge and his swollen limbs off the muddy floor, the other miners have built him a bed from a wooden pallet, and he lies there for hours on end, staring at the ceiling. Jorge is turning gray; they are all turning gray. Their faces and arms have lost the cinnamon and bronze of the South American sun. Instead, they are the hue of mushrooms, of watery ash.

The mushroom men avert their eyes from one another, as if they are ashamed of their appearance, though it's not vanity that causes them to do this. It's the way they feel inside: small, broken, like a dog that's been kicked, or a boy who's been teased so often he believes that humiliation is all he deserves.

On their seventeenth night underground, the men hear another drill getting closer. The *rat-a-tat-tat, grind-grind* is getting louder, holding the promise of either their liberation or another disappointment.

Víctor Segovia won't allow himself to believe the drill will break through. Instead, he asks Mario Sepúlveda: "What do you think dying is like?"

Mario says it's like falling asleep. Peaceful. You close your eyes, you rest. All your worries are over.

Up on Level 105, the sound of the drilling is getting closer as Raúl Bustos falls asleep, causing him to have an odd and hopeful dream. He's been thinking about his children, and especially his six-year-old daughter, María Paz. She's a bright and competitive little girl and is always winning races and trying to get perfect scores on her tests. In his dream she's operating the drill that's trying to reach him. "She always wants to win, she has a strong personality," Raúl says. He asks her, "Please, María Paz, come and get me. You can do it." She answers: "I'm going to win, Papá, I'm going to get you out of there." María Paz doesn't like to lose, and in his dream Raúl is hopeful that his daughter, a six-year-old girl driller, will reach him and save his life.

Sleeping nearby, Alex Vega dreams that he is climbing through the mine. He squeezes past the stone wall blocking the Ramp into the cavern of the Pit, and begins to crawl and grab his way upward over boulders, rising ever higher until he reaches the opening where the weathered old winch building stands. He walks out onto the surface and sees an entire city of rescuers and drillers trying to reach the men below. "We're alive, we're down there," he tells them. "I can show you the way."

The Terraservice team is getting closer. On the morning of August 21 their third attempt to reach the miners has drilled 540 meters deep: Their target is a gallery next to the Refuge, or the Refuge itself, at 694 meters, or 2,276 feet, beneath the surface. Nelson Flores is the drill operator working the day shift, which involves standing on a grill-shaped platform attached to the Schramm T685 truck, monitoring two gauges that measure the torque at which the drill is turning and the air pressure the truck is sending to the drill bit and hammer below. The diorite is good for drilling, Flores thinks, it's free of cracks, as smooth and even as gelatin. Every six meters they stop to add another steel tube to the drill assembly, and begin drilling again. Flores touches his hands to the shaft and raises a lever, increasing the pressure until he feels the hammer kicking in. As the drill goes lower, the hammer pulse transmitted to the surface through the steel shaft gets weaker and eventually Flores has to close his eyes and concentrate to feel it working. He drills until sundown and heads home, passing the entrance to the camp, where he and the other drillers who've finished their shifts hear another round of applause from the family members.

María Segovia is still at her tent, closest to the gate, with her family and the Vega family nearby, all about to begin a night of nervous anticipation. Golborne and Sougarret have informed the miners' relatives at the daily briefing that one of the drills is getting closer and there is a chance it could break through the next morning. Jessica Vega normally goes to sleep in her tent a bit after midnight, but tonight she'll stay up late into the night with a group of relatives that includes Alex's younger sister Priscilla and Priscilla's boyfriend, Roberto Ramirez, both of whom are just shy of thirty. The young couple are singers and Roberto has a

mariachi band (Mexican music is popular in northern Chile) and mariachi sideburns to match. He's brought his guitar, to liven things up,
and maybe to celebrate, because this might be the last night they spend
without word from Alex. Roberto can already feel it's "a special night, a
magical night." His spirits have been lifted, unexpectedly, by what he's
seen in the drive into the mine. A storm passed through the desert a few
days earlier, dropping a light, rare rain on the driest desert on Earth. The
average annual rainfall in Copiapó is less than half an inch, but this is
an El Niño year, and a somewhat early rain (the first storms in the Atacama usually come in September) has briefly moistened the surrounding land and produced what's known in Chile as the *desierto florido*,
the flowering desert. Roberto has seen the rocky and sandy landscape
of khaki- and copper-colored mountains covered, suddenly, with fields of
fuchsia, white star-shaped flowers, and yellow trumpets swaying in the
breeze. It's enough to make a man want to sing.

As night falls in the camp, the desert breeze blows at the bonfire the
Vega family builds next to their camp, and Roberto begins to strum his
guitar. The Segovias in the next set of tents are unusually quiet, and
Ramirez and the Vega family feel the need to fill the silence by making
a "racket." It's about two or three in the morning and they've been singing for an hour or two when Roberto tells Jessica that he's written a song
in honor of Alex. He pulls the lyrics from a piece of paper in his wallet.
Like many Latin American folk songs, it tells a true story, in this case
the history of the events the Vegas and the other families are living, and
it opens with a slow, sad tempo as it describes the mournful mood in the
Copiapó neighborhoods where many of the miners live.

Cuando camino por las calles de mi barrio
no veo el rostro feliz en los familiares.
En Balmaceda y Arturo Prat
sin ti no existe un mundo mejor.

When I walk through the streets of my barrio
I don't see happiness on the familiar faces around me.
In Balmaceda and Arturo Prat
there is no better world without you.

Next, the song describes the mountain's collapses, and José Vega's attempt to enter the mine and reach his son.

Se desintegran las rocas del cerro
Los mineros pronto saldrán
La chimenea está colapsada
Pero tu padre pronto te sacará.

The rocks of the mountain fall apart
The miners will soon come out
The chimney has collapsed
But your father will soon bring you out.

The chorus introduces Alex's family nickname—"El Pato," the Duck, a name given to him when he was a boy—and speeds up to the livelier tempo of a protest song. It's the kind of chanted tune you might hear a thousand people sing as they march through the streets of a Chilean city.

Y El Pato volverá!
Y va volver!
Los mineros libertad
Y va volver!
En el campo o en el mar
Y va volver!
Y también en la ciudad
Y va volver!

And El Pato will return
And he will return!
Liberty for the miners
And he will return!
In the countryside or in the sea
And he will return!
And also in the city
And he will return!

The singers ask Alex to come back to the home he was building with his wife, to that little property on the sloping street where husband and wife spent the weekends building a concrete wall together.

Pato vuelve a casa,
Tu esposa y tu familia te esperan
vuelve ya.

Y El Pato volverá,
Y va volver!

Pato come back home
Your wife and your family are waiting for you
Come back now.

And El Pato will return,
And he will return!

The Vegas and Roberto Ramirez sing the song several times, late into the night. Finally, they go to sleep, because the word from the government is that the drillers won't reach the level where the miners are until much later in the morning.

As the Vega family sings during the early morning hours of August 22, Mario Sepúlveda, overcome from several days of insomnia and agitation, slips into the deepest sleep he can remember. All the tension lifts from his mind and body. In the surreal vividness of a hunger-induced dream, he finds himself transported to that place where all of his longing, his hurt, and his love of life were born. He's in Parral, sleeping on the floor, and when he lifts his head he sees his grandmother Bristela and his grandfather Domingo, "all dressed up and beautiful." They've been dead for many years, and in his sleep Mario feels the joy of a man witnessing a resurrection. They were his *viejos*, the people who cared for him most as a small, motherless boy. His grandmother has brought a basket filled with food. *Porotitos con locro*: beans in a winter stew with corn and meat still on the bone. "Get up from there, *hombre*," his grand-father tells him, in the strong, country voice of an old man. "You are not going to die here." *Vos no vas a morir aquí.*

Nelson Flores, the drill operator, is home for just two hours or so before the call comes in ordering him back to work. The driller from the night shift has had a family tragedy—his grandmother died. So Nelson re-turns to the San José Mine and borehole number 10B. He works through the night, the sound of the drill drowning out the singing from the Vegas farther down the mountain. Just after 5:00 a.m., with the winter sun beginning to turn the horizon indigo, he has the drill bit advancing at just 6 to 8 meters per hour. He stops to allow the crews to add another 6-meter-long tube to the shaft. It's linked to 113 other tubes below it, the borehole 10B having now reached 684 meters, about 10 meters from the spot where they hope to break though. When the men are done, Flores closes his eyes as he raises the lever that adds air pressure to the ham-mering drill bit. The 114 interlinked tubes begin to turn, moving the drill bit at the bottom of the shaft and its tungsten carbide beads. Tung-sten carbide is harder than the granitelike diorite, and in the friction bat-tle between the two, tungsten carbide wins, grinding the diorite into dust that is shot by air pressure more than 2,200 feet to the surface, produc-ing a cloud of lead-colored dust the drillers call a "cyclone." With that pillar of dust shooting from its chimney, the Schramm T685 resembles some kind of stone-powered train, and Flores's boss, the drill supervisor Eduardo Hurtado, sits nearby in a pickup truck and watches as the sil-houette of the cyclone rises steadily against the sky, a sign that the drill under their feet is advancing as it should.

Sometime after 5:00 a.m., Mario Sepúlveda wakes up at the command of his dead grandfather, and the good, almost euphoric feeling of a dream come to life stays with him in the minutes that follow, as he takes in the grinding and pounding sound of a drill that's become impossibly loud.

Richard Villarroel, the expectant father, has been trying to sleep. He's about forty-five vertical feet above Mario, in the passenger seat of a pickup truck, at Level 105. The sound of the drill approaching is very loud, yes, but there's no real way to tell if it's actually as close to breaking through as Richard hopes it is. He's been reciting the Our Father and Hail Mary over and over again, about one hundred times in all, with assorted pleas to Jesus. When the pounding stops, briefly, at 5:00 a.m. he says, "*Papito* [little father], help that operator change the bars [that house the drill], and guide him to us, please . . ." He still can't sleep, so he goes down to the Refuge, where there's an insomniacs' game of dominoes in progress, with the set Luis Urzúa made. Richard joins a match with José Ojeda, a bald, short man and mine veteran. After a while, the drilling starts to get even louder.

"It's going to burst through," José says in a matter-of-fact voice.

At about 6:00 a.m., several of the men around the drill operator, Nelson Flores, have fallen asleep. No one expects a breakthrough for several hours. Flores notices something odd: The last steel tube turning the 114 tubes below is starting to stutter in its rotation. The drill bit is grinding away at something with a different texture. Suddenly the cloud of dust coming from the Schramm's chimney stops, and the pressure gauge drops to zero. Instinctively and immediately, Flores lowers a lever that shifts the drill engine into neutral and stops the air pressure being forced down into the shaft. As he does, the rig turns quiet, and the sudden silence is filled, almost immediately, by the sound of his boss and coworkers yelling and running toward him.

Far below, 688 meters under Flores's feet, there's a small explosion just up the tunnel from the men in the Refuge—*poom!*—followed by the sound of rocks tumbling to the ground. The grinding of metal against rock that has filled the ears of the men stops, and in its place there is a whistle of escaping air. Richard Villarroel and José Ojeda jump up and run toward the noise, Richard grabbing his 48-millimeter wrench as he goes. They are the first to reach the spot. A length of pipe is protruding from the rock, at the spot where the wall and ceiling meet, and Richard watches as a drill bit inside the pipe lowers and rises, and lowers again:

Up on the surface Nelson Flores realizes he's entered an empty space, and is "cleaning" the shaft. Then the drill bit falls to the mine floor and stays there, and Richard takes his wrench and begins pounding on the exposed pipe protruding from the tunnel ceiling.

Richard has been waiting for days to put this wrench to use. It's two feet long, the biggest chrome-vanadium tool in his possession, and now he strikes it against the exposed pipe with joy and desperation, a repetitive clank that's meant to announce a human presence to the drill operator above: *We're here! We're here!* He strikes the pipe, and the idea that he will see his first son born takes hold of him, that his prayers have been answered by this drill bit and the men who sent it here. Richard pounds away until his boss, Juan Carlos Aguilar, steps in behind him and tells him to stop, because they have to think like miners, and reinforce the roof of the tunnel where the drill has broken through, to keep from being crushed by a loosened slab of rock.

Soon all thirty-three men have gathered around the pipe and the drill bit, objects that have intruded into their dark world with the promise of raising them up to the world of light again. With its parallel circles of pearl-size tungsten carbide teeth, the drill bit resembles some Assyrian sculpture, a kind of alien apparition, and the men stare at it in awe and joy, embracing and weeping. To Carlos Mamani, who falls to his knees before the drill bit, "it felt like a hand had punched through the rock and reached out to us."

José Henríquez, the jumbo operator who's been transformed underground into a shirtless and starving prophet, looks at the drill bit and pronounces the obvious to anyone who will listen:

"*Dios existe*," he says. God exists.

PART II

SEEING

THE DEVIL

10

THE SPEED OF SOUND

For the first few minutes after the breakthrough of drill 10B, the men keep pounding at its shaft. They take turns, hitting it not just with Richard Villarroel's chrome wrench, but also with loose stones and a hammer, not paying much heed to those warning that the rock loosened by the drill could fall on their heads. "We were like little kids hitting a piñata," Omar Reygadas remembers. *Como cabros chicos pegándole a una piñata.* The shirtless boys in the yellow, blue, and red helmets keep hitting, until one of the miners drives up with a forklift, which lifts up Yonni Barrios and Carlos Barrios in a basket to perform the critical task of reinforcing the spot with steel bars. They're frantic, yelling instructions back and forth: Above all, they have to erase any doubt the people on the surface might have about men being alive down here. Make a sound, leave a mark, attach a note. Someone says to stop hitting the bar, to see if the people who are at the top are answering, and Yonni puts his ear to the bar and says, yes, he hears them tapping back. A miner tosses Yonni a can of red spray paint, and he tries to leave a mark on the shaft, but the steel is covered in a stream of muddy water flowing from up above that erases the paint again and again. "We needed to dry the bar, but we didn't have anything dry to clean it with." Eventually some of the paint seems to stick. The men tie the notes and letters they've prepared, more than a dozen in all, wrapping them in pieces of plastic and strips of electrical tape and rubber tubing against the moisture that's pouring down through that hole, worrying that a piece of paper might not survive the long journey back up through the slosh. They keep pounding on the bar.

Nelson Flores, the drill operator, feels the pulse in the steel from down below before he hears it. At first he wonders if it's just the weight of the 115 steel bars, 22 tons' worth, striking and settling against one another in the shaft. Putting his ear to the uppermost bar in the shaft, he hears a tapping that's fast and frantic, but which then slows, "as if the *viejos* down there were getting tired." As word goes out for all the other drills on the mountain to stop, several other men listen to the sounds coming from the steel tube. "It's them!" The drill team moves quickly and carefully to add one more steel bar to the shaft, so they can measure how deep the cavity is by lowering the bit until it strikes something and stops. When they're done, Flores watches as the shaft moves four meters before it stops, which is exactly the height of the passageway they were aiming for. Listening to the shaft again, they hear that the sound from below is shifting in rhythm: It begins tapping out as if sending a Morse code signal, or making music, mixing short and long pauses. "At that point, we had no doubt," says Eduardo Hurtado, the drill supervisor. "There was someone alive down there."

Calls go out to various Chilean officials. Sougarret, the engineer in charge of the rescue, is skeptical. He issues an order to the drill team that will be immediately disobeyed: "I told them not to tell anyone, because I remembered what happened the last time we broke through. I didn't want to cause another crisis with the families." Minister Golborne is cautious, too, and since it's still before 6:00 a.m. and President Piñera in Santiago is likely asleep, Golborne sends his commander in chief a text message: "*Rompimos.*" We broke through. There is to be no news, officially, to the families or anyone else, but after so many days of frustration the drillers can't contain themselves, and word starts to spread among the dozens of rescuers and support staff gathered inside the mine property. Pablo Ramirez, the friend of Florencio Avalos and the man who first entered the mine in search of the trapped men with Carlos Pinilla, hears the news and rushes to the site of drill 10B. Many of the rescuers know Ramirez by now, because he's been consulted time and again for his knowledge of the mine, and they know he has many friends buried down there, and when he arrives at the drill they allow him to listen. The sound coming from below is louder: It's clearly and undeniably human, even after traveling 2,200 feet through steel. The trapped miners are so far away that if they were simply yelling into that shaft it would take more than two seconds for the sound of their voices to travel through the air to reach the top. But sound moves through metal twenty times faster, so Ramirez can hear within a fraction of a second each time his friends below strike the drill bit.

There is now cell phone service at the mine site, thanks to the efforts of the Chilean government, and Ramirez calls the first person he wants to know, Florencio Avalos's teenage son, Ale. It's Sunday morning, and for once Ale isn't rushing back and forth from the mine to the school.

"Ale, your father is safe," Ramirez says. "Don't worry. They're all alive. Listen." Ramirez places the phone against the steel shaft.

At home in Copiapó, Ale hears the sound coming from the place where his father is buried alive.

"It was like a bell," Ale remembers. "Like a bell you hear at school."

Ale calls Camp Esperanza, where his mother is inside her tent, having drifted off to sleep an hour or so earlier.

"Mamá, Uncle Pablo says they're all alive."

Mónica gives thanks to God, and "only to God," and there is something defiant in the way she says this, because she realizes at that moment how alone she's been since the night of August 5. "It felt like my heart had opened up again." Florencio is alive and her life is going to start over. After seventeen days of eating only sparingly, of briefly forgetting her own children, of sleeplessness, hunger, and sleepwalking, Mónica will once again start to cook and eat on a regular schedule. When she steps out of her tent and into Camp Esperanza, she sees her in-laws at their own tent nearby. She wants to tell them the good news, but before she can open her mouth it's clear they already know. While she was sleeping, a few rescuers have come running down from the drill site yelling: "We found them!" Those words have reached the ears of her in-laws, who haven't thought to wake her up and tell her. Since their son Florencio was trapped they've kept their distance from their daughter-in-law, they've watched as Mónica fell apart, and did not, or could not, help her. They've seemed angry with her, and perhaps it has something to do with the fear that their undeniably bright son has been killed working in a mine to support the family that he and his pregnant girlfriend started when they were only fifteen. Mónica is confused and hurt. Her moment of joy is mixed up with this new family wound she never expected.

Mónica and her in-laws look at each other for an awkward moment. "No *importa*," she says. It doesn't matter.

In Copiapó and across Camp Esperanza and the ground underneath, the drama and the longing surrounding the fate of the thirty-three men has been entangled with the messy complications of everyday family life since August 5. This hopeful morning of August 22 is no different. Susana Valenzuela shares the news with Marta, her boyfriend Yonni's wife, and in other families siblings and cousins who spent years avoiding each other are joined together, again, by the sudden wonderful possibility that the man whose love they all seek, whose life they've prayed for, might be alive after all. It isn't easy being a miner's wife, or his girlfriend, or his son, or his daughter, or his ex-wife. Before the accident Darío Segovia's adult children from his previous marriage had never talked to Jessica, his current love and the mother of his baby girl. With Darío buried underground, Jessica has met two of these adult children for the first time, and for seventeen days the two hitherto separated halves of Darío's family have been thrown together by worry and the possibility of imminent, permanent loss. But the old resentments haven't disappeared. "I was never married to their father," Jessica says of Darío's adult children. "And sometimes I think they didn't want me at the camp." Her love with Darío was as real as the home they shared, as real as that last, long embrace he gave her before going to work, and perhaps Darío's older children have seen this love in the way Jessica waits with their half sister at the camp. Or perhaps they think she's "just one more woman" on Darío's "list" of conquests, as Jessica puts it. For one moment of happiness, none of that will matter—and then it will matter again. The thirty-three men are alive—it has not yet been confirmed, but that's what many at Camp Esperanza believe already—and they will return from the shift they started seventeen days ago. And everything in their aboveground lives is going to remain as complicated as it was before.

Mónica Avalos begins walking around the camp amid the embracing siblings, spouses, cousins, and children of miners. Many a prayer is heard, and it will be a day of piety and thanks in the camp some have called a "Jerusalem." Once, Mónica strode across the dry, dusty surface of the mountain in her sleep. Though they are filled with tears, her eyes are open this morning and she is fully awake and alert and present for the first time in seventeen days, watching the camp and its wives and girlfriends and brothers and sons speak, their breath visible in the air of a morning just turning to light.

The drill bit with its tungsten carbide beads spends four hours on the floor of the passageway above the Refuge before it begins to rise up into the 4.5-inch shaft through which it came. The thirty-three men watch from a safe distance as it disappears into the hole, carrying their messages: a few personal letters, details about where, precisely, the drill has broken through, and one very pithy note written by José Ojeda, who's condensed the most critical information (how many of them are alive, their physical condition, and their location) into just seven words written in big red letters. He's wrapped his note behind the hammer, because one of the miners said that would be the safest place. The thirty-three men gather around to celebrate, Mario Sepúlveda calling the last stragglers to a gathering by the Refuge, "Florencio, Illanes, get over here!" They start to chant "¡Chi-chi-chi, le-le-le, mineros de Chile!" José "the Pastor" Henríquez has turned on his cell phone camera to record the moment. More than half of them are stripped down to their underwear against the heat, and they look like a group of homeless men who've decided to stage a scene from that novel and film about castaway boys, Lord of the Flies. They are laughing and cheering, and passing around a plastic bottle filled with dirty water as if it were champagne. The haunted, concentration-camp look that covered their faces just a few hours earlier is gone. Mario Sepúlveda throws up his hands and makes the aggressive, pleading gesture Chilean men make at soccer matches. Alex Vega wraps his arm around him and soon the entire group launches into a rendition of the national anthem. As they begin to sing they are shouting the lyrics, especially the first lines about Chile being a "happy copy of Eden," though by the time they get to the third repetition of the final line about "the refuge against oppression," their unfed voices have started to sound meek, and the song peters out.

When Minister Golborne arrives at the mine, he heads for Camp Esperanza first, before going to the drill site. He officially informs the families of the news they already know: that they've broken through and are hearing sounds from below. He finds María Segovia and the others and promises that as soon as the rescuers confirm that the miners are indeed alive they will be the first to know. By all accounts, Golborne has worked hard in the previous days to win over the trust of the family members; a few days before, they'd given him a miner's helmet signed by the families, and it will soon become the most precious memento of his eventful days there. *Vamos, Ministro, déle con fuerza. Confiamos en Usted*, the helmet says. Go, Minister, give it all you've got. We trust you. The minister then heads up to drill 10B, where Hurtado and his team have a stethoscope for him to listen to. What the minister hears certainly sounds like men tapping at the shaft, but when he calls the president he is cautious. "I can't be completely certain. It could be the power of suggestion."

The president is in Santiago, and he also talks via phone to Cristián Barra, his fixer at the Ministry of the Interior. "Should I come?" the head of state asks. Barra tells the president to stay in Santiago, because it's possible only some of the men are alive, and that the government will have to make a grim announcement about how many are dead. But it was only a rhetorical question, because the president is already in his car, on the way to the airport for the hour-long flight to Copiapó.

As the president heads north, the drill team is slowly raising up the bit and removing the 115 four-hundred-pound steel tubes from the shaft, one at a time, a process that will take up the rest of the morning and part of the afternoon. President Piñera is still en route as the Terraservice team prepares to remove the last of the steel tubes inside the shaft and the hammer and drill bit attached to it. Only a few workers and officials are allowed at the site, though dozens more are hovering nearby, outside the security cordon Barra has placed around the shaft.

Barra has ordered that no one is to leave the area, lest some bad news filter out before the government can make an official announcement to the hundreds of people gathered in the camp below. By now it's a bright, sunny, chilly afternoon of South American winter, and Golborne and the other Chilean officials are wearing both sunglasses and red government jackets. Finally, the last tube emerges from the shaft, covered in mud. The drillers pour water and wash away the muck, revealing a clear red mark on the metal: The miners painted several feet of the steel tubing, but only a single, palm-size smudge has survived the journey through stone and mud to the top. "Was that there before?" the minister asks. "No!" comes the excited reply from the drillers. They've found confirmation that at least one man is alive down below, and many of the men gathered around drill 10B exchange quiet embraces. Golborne can see there's something wrapped around the bit, and he begins to remove it. It's some sort of rubber tubing, and he lets it fall to the ground, because underneath the tubing a piece of paper is visible. Of the dozen or more notes the men attached to the drill assembly, three have survived, and Golborne has just found the first, removing the paper carefully, because it's wet and it immediately starts falling apart in his hands. "No, don't unfold it, Señor Ministro," someone says. "Wait until it dries." "If we don't read it now, we'll never be able to read it," someone else says. Finally, Golborne gets the first note open.

"What does it say?"

The minister of mining begins to read out loud: "The drill broke through at Level 94, at three meters from the front. On one side of the roof, close to the right wall. Some water is falling. We are in the Refuge. Drills have passed behind us . . ." Part of the note is cut off. It ends with: "May God illuminate you. A *saludo* to Clara and my family. Mario Gómez."

Barra begins to read a second note: "Dear Lilia. I am well. I hope to see you soon . . ."

"It's a personal letter," someone says. "We should save it."

As two of the more powerful men in Chile are trying to decipher these messages, one of the sunburned roustabout members of the mining crew has quietly used his feet to move the piece of rubber tubing Golborne first tossed on the ground. The driller figures he'll hold on to it as a souvenir, but when he begins to take a closer look at what he's going to take home, he notices there's something hidden inside. "It's another note," someone next to him yells, and soon the minister himself is opening this third message, written on a folded piece of graph paper.

ESTAMOS BIEN EN EL REFUGIO. LOS 33.
WE ARE WELL IN THE REFUGE. THE 33.

Even before Golborne can announce to the men what it says, those looking over his shoulder scream out in joy. *¡Vivos!* Each and every one of those knuckleheads down there is alive. *¡Todos los huevones!* Suddenly all the workers are cheering and embracing, and one of the drillers falls to his knees. Some of the men embrace again, but a few begin sobbing as they do so, bawling the way men do when their mothers die, or when their sons are born. These rugged men have been sending steel bits into the gray rock beneath their feet, and they are at this moment surrounded by piles of that rock, and the dust made from boring down into it. They're men who drill holes looking for gold and copper and other metal, and they've drilled the greatest hole in their lives to reach thirty-three knuckleheads and find them, under this seemingly immovable, unconquerable mountain.

"*Gracias, huevón, gracias.*"
¡Lo logramos! "We did it!"

The burst of triumphant emotion makes everyone forget the Ministry of the Interior's security "protocols," and no one moves to stop several drillers as they run downhill, away from the Schramm T685, toward the fence that separates the mine proper from Camp Esperanza, toward the tents and the shrine and the kitchen there, and the television satellite dishes, and the cords of firewood, and the lines of smoke climbing up from the recently extinguished campfires, including the one where Alex Vega's family and friends stayed up late into the night singing a ballad to him. The rule-breaking drillers shout, loud enough to be heard back up at drill site 10B, because now all the drills have stopped and all the machines that have filled the mountain with machine noises are silent, and the mountain is covered instead by human sounds, a cheering that's spreading, with the cries of the drillers loudest.

"All those bastards are alive! Alive! All of them!" *¡Están todos los huevones vivos!*

A short while later the helicopter that is transporting the president from the Copiapó airport to the mine lands nearby. The families and the media gather before him, to once again hear officially what everyone on the mountain already knows. The president has the honor of showing José Ojeda's note publicly for the first time, with its bold red letters as proof to any skeptics that the improbable is true, an image that when broadcast sets off celebrations across Chile. From the northern border town of Arica, where Víctor Zamora, the miner who is always hungry, lived in a children's shelter, to the Patagonian towns halfway to Antarctica, where the soldier Juan Illanes passed through one Christmas, there are cheers and shouts as people run from their televisions to the streets and plazas. In Copiapó the discovery of the thirty-three men is celebrated with peals of church bells, each percussive collision of metal against metal sending sound waves racing through the Sunday winter air.

11

CHRISTMAS

A camera, speaker, and microphone are lowered into borehole 10B, as the president of Chile, Minister Golborne, and assorted officials watch. A psychologist, Alberto Iturra, is also present, and he's deeply concerned about what condition the men will be in after seventeen days buried alive, since according to the government's best (and private) statistical calculations, they should be dead. They are almost certainly suffering from some sort of state of altered consciousness, and Iturra is irritated because his advice to the leaders of the rescue team has been ignored. The psychologist thinks that the first voice the miners hear from the surface should be a familiar one, and he's suggested that it be Pablo Ramirez, Florencio Avalos's compadre and a friend to many of the thirty-three trapped men below. But the officials present have overruled the psychologist, because the president is there and wants to speak to the miners in the name of the Chilean people and who can say no to a president? The miners are safe and the whole world is tuning in to the great miracle of the San José Mine, and a few rays of the miracle's holy glow are going to shine on the recently elected Piñera. Now that the story is a happy one, and not a tragedy, a bit more politics and vanity will mix with the undeniably abundant altruism and selflessness so far displayed by the rescuers and officials at the San José. "We started to have a lot of issues with egos and flags," the psychologist Iturra says. Take, for example, the very camera, speaker, and microphone now descending toward the trapped men below. The Chilean navy and Codelco have just had a small bureaucratic kerfuffle over which arm of the government would send down that equipment and provide the operators to control it. The navy has some excellent cameras it uses for submarine rescues, but Codelco has its own technology, and in the end it's clear that Codelco "owns" the hole, says Iturra, who adds wryly: "But Codelco didn't own the miners: The miners belonged to the Chilean [social] security administration." The middle-aged psychologist, who is a bit vain himself (he volunteers the fact that he was a math prodigy as a young man, and also an engineer), thinks that he should be there at the microphone, too, or nearby at least. But he's in the background as the Codelco camera descends into the shaft, transmitting to a screen on the surface the image of an endless tube carved into the gray diorite, its edges looking moist and fleshy, as if the camera were probing the entrails of some great stone beast. The camera reaches the bottom and the image shifts out of focus and into darkness.

Darío Segovia, Pablo Rojas, and Ariel Ticona are keeping watch at the hole, babysitting the steady stream of water passing through the opening that leads to the surface. They are waiting to see what comes down next, and after a long time they finally see a gray electric light falling toward them, growing in intensity, and they begin to shout.

"Something's coming! Hurry!"

All thirty-three men gather around the hole. They see a glass eye, on a kind of swivel. Luis Urzúa thinks it's a mining scanner of the kind he's seen used in exploration before, but then another miner says the obvious: "It's a camera."

"Lucho, you're the boss, speak to it! Show yourself!"

Urzúa walks up close to the camera. He wonders if it has audio attached to it. (It does, but it isn't working: Unbeknownst to Urzúa, the president of Chile is on the surface, speaking into a microphone.) "If you can hear me, move the camera up and down," Urzúa says. The camera begins to move—in a circle. Urzúa follows it, doing a funny circle dance, until it catches his eyes and stops.

Up on the surface, the president, André Sougarret, and a team of officials and technicians look as two eyes stare back at them on a black-and-white screen, looking eerily neutral, dreamlike. The psychologist,

Iturra, sees those eyes and a light above it, and then more lights attached to the helmets of men moving in the background. Seven lights in all. And he thinks: *Well, there are at least seven men down there who can help us take care of the other twenty-six if it comes to that.*

Down below, the joy of being found alive is starting to wear off quickly. "We are very hungry," Víctor Segovia writes in his diary. "The mountain cracks and rumbles a lot." Several hours pass as the rescuers above work on the shaft. "There are a lot of arguments. The mood is very bad." Mario Sepúlveda gets in an argument with Juan Carlos Aguilar's contract mechanics: A "misunderstanding," Segovia writes. All the miners start talking about what kind of food they'll get first. A Coca-Cola, a chocolate bar, perhaps. What can fit in that tube? A beer! Many delicious and filling foods and beverages can fit in a tube 4.5 inches wide, but for the moment the orifice exudes nothing more than dripping, dirty water—there's so much water, they're going to have to fashion a gutter to drain it off. "What's going on up there? Why are they taking so long to give us food?" Finally, at 2:30 p.m., more than thirty-two hours after the drill first broke through, and more than eighteen days since most of them have eaten a meal worthy of the name, another object comes down the hole. It's an orange-colored PVC tube, sealed off with something inside, like some oversize and stretched-out plastic Easter egg. The tube has a wire dangling from it. "There's a wire attached to this, get Edison," someone yells, because Edison Peña is an electrician. Edison opens the tube, and he sees there's a telephone wire inside, and a handset receiver.

There are all sorts of experts and technicians in the army of rescuers, and the Chilean government at this point is getting advice from around the world, including from NASA. But the phone that's been lowered down into the tube is a secondhand instrument, fashioned from recycled phone parts by a thirty-eight-year-old Copiapó businessman. Pedro Gallo runs a communications company that serves the local mining industry, and he's been hanging around the mining site since August 6. He's been trying to volunteer his know-how, but to some he's become a bit of a pest, and certain rescue officials with Codelco have told him he's not allowed on the site. Gallo has no relatives inside the mine: He's there, like dozens of others, because he senses that an epic story is unfolding on the windswept mountain. Gallo has stayed at the mine, hoping to play a role in the drama, even though his wife has been calling him to come home—she's seven months pregnant. Finally, on the morning of August 23, he gets his chance. "We need you to make that phone line you've been talking about," a Codelco official tells him. In forty-five minutes, using some old telephone components, a piece of plastic molding, and a few thousand feet of discarded wire, he's put together a telephone receiver and transmitter.

At 12:45 p.m., with Minister Golborne and Carlos Barra and assorted other people watching, Gallo's telephone is lowered down toward the trapped men. The wire attached to the receiver is composed of nine separate stretches of phone line linked together rather crudely, with knots and electrical tape, and at one point the minister asks: "What are those things, are they transmitters?" Gallo answers: "No, Señor Ministro, that's where I patched the wires together." After fifty minutes, the handset has been lowered 703 meters and has reached the bottom. Gallo connects the final stretch of wire to the kind of cheap phone set you can find in a million offices around the world.

Golborne lifts the receiver and speaks a few words of mining protocol suggested by the phone technician.

"Attention, mine shift," the minister says. "The surface here."

"Mine, here," Edison Peña answers. "Can you hear me?"

"Yes, I hear you," the minister says, and when he does the two dozen men clustered around the telephone on the surface burst into cheers and applause.

Down at the bottom of the shaft, Edison can hear what's happening on the surface with unexpected clarity, the earpiece filling with the robust, hopeful voices of living people in the outside world. "I could hear this collection of people. [*La colectividad de esta gente*]. And I heard this very firm voice . . . I broke down." After eighteen days in darkness, after the hours of silence when he lived with death and the thought that no one was coming for him, Edison is overcome by emotion. The sound of those strangers' voices makes him weep. "I just wasn't capable of speaking."

"This is the minister of mining," the minister says.

Someone takes the phone from Edison and says they will pass it to the shift supervisor. "Yes, with the *jefe de turno*, that's the right thing to do," the minister says, and the minister turns on the speakerphone so that everyone around him on the surface can hear.

"It's the shift supervisor, Luis Urzúa."

"There are twenty of us here ready to provide you with immediate help," the minister says. "How are you? How do you find yourselves?"

"Well. We are well. In good spirits, waiting for you to rescue us," Urzúa answers, though his voice is hurried and uncertain.

The minister says the rescuers will soon be sending down drinking water, and some liquids with instructions from a doctor.

"We've been drinking some water," Urzúa says. "But at this moment we've already eaten the little we had in the, the, the, the Refuge."

The minister says he will soon put them in touch with a doctor who will be in charge of feeding them. The men below are excited, and desperate to get out, but no explanation of how and when they will be rescued takes place in this first conversation. Instead Golborne, swept up in the emotion of the moment, feels the need to give the miners a sense of how much their survival has meant to the Chilean people. "I want to tell you the entire country has been following you these last seventeen days. The entire country has participated in this rescue," the minister says. "Yesterday all of Chile celebrated. In all of the plazas, in all the corners of this country, people celebrated that we'd made contact with you."

Now it's the miners down below who start cheering, the sound of their yells and their applause tinny on the speakerphone. For men who are, at this moment, living half-naked and half-starved underground, this bit of news from the minister carries the aura of the fantastic. While they were inside this rocky tomb, alone in the darkness, an entire country was out there praying for them, thinking about them, working to get them out. It's as if they've stepped out from a dark grave into the magical glow of a fairy tale.

As the cheering dies down several miners begin gesturing at Urzúa. They want the shift supervisor to ask about Raúl Villegas, the driver who was headed out at the moment of the collapse.

"Can I ask a question?" Urzúa says into the telephone.

"*Sí*," the minister answers.

"We had a colleague who was headed outside. A driver," Urzúa says. "We don't know if he made it out."

"Everyone made it out unharmed," the minister says. "There is not one injury or death to lament."

The thirty-three men begin cheering again. Another element of the miracle of the San José Mine has fallen into place, and now the minister reveals one more. "There's a camp out here with all your families," the minister says. Their families have been waiting and praying for them, the minister adds, and for the thirty-three men it's as if a veil of solitude and hurt has been lifted: The people who've loved them are on the surface, directly above them, gathered around the hole they entered to work eighteen days earlier.

A bit later, the rescue leader, André Sougarret, comes on the phone and tells the miners to stay away from the rock that's blocking the Ramp and the chimneys to the surface. "Because it could keep falling," Urzúa says. "Correct," Sougarret says. Cristián Barra, the president's fixer, comes on the line to say: "I send you a greeting from the president. He's been here four times already." Not long ago they were nobodies, but now the president is sending them *saludos*. Finally the first phone call between the trapped miners and the surface ends with the miners singing the national anthem. A government video captures the rescuers listening to their singing voices on the speakerphone. Later that day the video is released to the global media, along with the telegenic image of Minister Laurence Golborne, in his official red jacket, beaming as he listens to the voice of Luis Urzúa. On many newscasts around the world, a photograph of Urzúa accompanies the sound of his voice, and he's identified as the miners' "leader." But who is really in charge down below? Iturra, the psychologist, prepares to ask each miner that question privately, even as the rescuers begin to lower down the first sustenance to the trapped thirty-three men.

What arrives in the next tube lowered down to the men is not a feast, or even anything that can be chewed. Instead the men receive thirty-three clear bottles with a few ounces of glucose gel. Not everyone is strong enough, at first, to help unload this precious cargo. "They'd go off to sleep, because they were so weak," Yonni Barrios remembers. Yonni unloads the tube with Claudio Acuña, José Ojeda, and Florencio Avalos. The rescuers have also included a set of instructions warning the men not to drink the gel too quickly, but of course almost all the men swallow it in one gulp, and several soon begin to feel their stomachs cramping painfully. When is the real food coming, the men want to know. Another tube comes down, but it doesn't have food either, but instead a form to fill out. The Chilean government wants each of the trapped men to provide his vital statistics (height, weight, age, shoe size, previous illnesses) and also to answer a series of questions about his current physical state: "When was the last time you ate? Are you urinating?" Most important, the bureaucracy in whose care the men have now fallen—and Chile's bureaucracy is the most relentlessly efficient in Latin America—asks that the men provide their R.U.T. number, the tax identification number that also serves as each Chilean's national identity number from birth. "Of course we had to give them our R.U.T.," Juan Illanes observes wryly. "They had to make sure it was really us down there." Without an R.U.T. you don't exist in Chile, and even Mamani the Bolivian immigrant has one.

Near the bottom of the form is a question included at the insistence of the psychologist Iturra. "*¿Quién la lleva?*" it asks, which translates loosely as "Who's running things?"

"We very specifically didn't ask, 'Who's the boss?'" Iturra says. Everyone already knew who the boss was down there, formally speaking. But was the "boss" really running things?

Juan Illanes looks at this question and is a bit perplexed. So are several other men. What do we put down here, a few miners ask him, because Illanes has been giving them these underground talks about the law and probably knows. Is Mario Sepúlveda running things? The third or fourth day after the collapse, Víctor Zamora had openly suggested that Mario be appointed leader in place of Luis Urzúa, only to be contradicted by Juan Carlos Aguilar. The contract mechanics follow the orders of Juan Carlos Aguilar: Should they write his name? Or maybe they should write Florencio Avalos, who's earned the respect of everyone with his energy and his quiet confidence. Truly, if they think about it, no one person is running things, they all are. But Illanes tells the people who ask him: "Put down Luis Urzúa. He's the boss." Illanes says this even though "Don Lucho's authority, at that moment, was hanging by a thread. If we [the contract mechanics] hadn't stood behind him, Mario Sepúlveda would have pushed him aside." Formal authority is a powerful concept in the life of a Chilean working man, and eventually most of the thirty-three men fill in the blank next to the question "¿Quién la lleva?" the same way: They write "Luis Urzúa." (Juan Carlos Aguilar, however, answers: "Everyone.")

As it happens, at that moment Luis Urzúa is beginning to take charge of one critical, technical aspect of the rescue. He's sitting in the front seat of his pickup truck, which has always been a kind of moving office to him, writing notes on paper. He is preparing to guide more drillers down to their location, because he can hear at least one more drill coming toward them. The rescuers will need a good, accurate map of the now-broken mine to reach the trapped men, and making such a map will require a new set of precise measurements underground. Luis prepares this information for the rescuers, working efficiently and competently today and in the days to come without making a big deal about it. But he's never been and cannot be the one and only "leader" who can corral and direct the egos and ids of thirty-two other workingmen trapped underground. That job would be tough for anyone, and it's about to get substantially more complicated thanks to events unfolding that afternoon up on the surface, where a black Hummer is rolling toward the San José Mine.

Leonardo Farkas is a Chilean dandy, and when he arrives at the mine it's in full sartorial splendor: stepping out from his Hummer in a long, double-breasted charcoal suit with a sky-blue tie and matching handkerchief, and stiff French cuffs with gleaming cuff links and assorted other jewelry that dangles and sparkles from his wrists as he speaks. He is a fit man, and his long blond hair dangles and catches the light, too, completing an odd and distinctive appearance, as if he were some Greek god reincarnated as a South American entrepreneur. Farkas is a multimillionaire whose investments include a nearby mine. He's also a fixture on the Chilean television charity circuit and he's come to the San José to dole out the kind of happiness only a carefree, wealthy man can provide: He's donating 5 million Chilean pesos (about $10,000) to each and every miner. The arrival of Farkas's black Hummer is covered live on Chilean TV. His subsequent conference with the miners' family members is an ostensibly private affair, though Farkas will later post a video of it on his YouTube channel.

After his assistants have already passed out slips of paper with the magic figure, roughly equal to a year's salary for the average Chilean laborer, Farkas takes to a small stage. A few family members start to chant his name. "Farkas! Farkas!" "I need the name of the person, the R.U.T., the bank account," Farkas says. "Those of you who don't have a bank account, the Bank of the State can open one for you for free." His big gift is just the beginning, he says. "Let's have every Chilean start contributing now. Each person can give one thousand pesos, five thousand, ten thousand." Farkas often donates large sums to the annual Chilean telethon that raises money for children with cerebral palsy and other developmental disabilities, and he talks about the trapped miners as if they, too, were needy children. His goal, he says, is to raise one million dollars for every one of the trapped thirty-three men. "Let's dream big. I've always dreamed big since I was a kid. Let's hope that before they get out, each one has a million dollars in their account." Farkas is a man who clearly enjoys the mass love his money can buy him, and this afternoon a few of the many waves of the joy and good feeling flowing over the San José Mine will envelope him, too. *"¡Gracias, Señor Farkas!"* But almost immediately that enormous sum of money, along with smaller sums donated by other people, and the promise of millions more, causes problems for the trapped men and their families.

For some families there's the problem that extra cash has always brought them when they're lucky enough to have it: Who gets to spend it? Who gets the luxury of being the object of the miners' monetary ministrations? There are several miners who are married and separated from their wives but not divorced. Chile was the last country in the Western Hemisphere to legalize divorce, just five years earlier, and most Chilean working stiffs haven't yet learned that they can and should pay a lawyer to end their marriages. Now that the de facto divorced miner is suddenly a millionaire (in the local currency), will his legal but unloved wife take control of this newfound wealth while he's trapped underground, or will it be his new domestic partner and their children?

Darío Segovia doesn't yet know about his *millionario* windfall. If he did, he'd start planning for the future, or paying off bills. For the moment his partner, Jessica Chilla, has decided she doesn't want anything to do with so much money: She asks Darío's brother to handle the family's Farkas cash. She can sense that the money is going to pit opposite sides of Darío's family against each other. The truth is, even before Darío and the thirty-two others had been found alive, when many people in his extended family took Darío for dead, "a lot of people just saw the peso sign in all of this," Jessica tells me. "The miners' lives didn't belong to them anymore, it was just about how much they were worth." Before the miraculous note surfaced, a few family members were privately making the same calculations of death benefits and insurance payments on the surface that Juan Illanes was making below. Now that the men are alive, each with 5 million Farkas pesos in their name, all the need that surrounds the people they've loved and brought into the world can be expressed openly and directly. *Before he loved you, I suffered alongside him . . . I was his son before he even met you . . . Don't we need to be taken care of, too?* "With all that money, the chicken coop [of relatives] gets all mixed up and the family gets warped," Jessica says.

In the hours that follow, the family members of the thirty-three men begin to write their first letters to the men down below. Several, like Jessica Chilla, decide that the issue of money is something her husband doesn't need to know about at the moment. Talking about money feels like tempting fate, or mocking God while their lives are in danger, trapped two thousand feet below the surface. But others, like Carlos Mamani's wife, Veronica Quispe, can't resist. They have two small children and in their lives as immigrants in Chile they've always struggled for money. Suddenly that cash burden has been lifted, that perpetual worry that sent the men into the mine in the first place. It's happy news, so in one of her first letters to her husband she writes: Thank God, Carlos, you are alive. After many days here on the surface waiting for word about you, we are happy and well. And one more thing. Thanks to Leonardo Farkas, we're millionaires.

Later that night of August 23, a second drill approaches the tunnels where the men are trapped. The rescuers on the surface have asked Urzúa to measure and map where the first drill broke through, but he already has that information available: It's 7 meters from one of the survey markers in the mine, number A40. The next drill, they tell him, should hit within 1.5 meters (5 feet) of the second one. The rescuers also ask Urzúa for more details on the physical condition of the men. He says that there are some who are very skinny, and that they're all very weak, but that there are no serious injuries. The doctors tell him to stop his men from drinking the dirty water down below—"we'll send you all the clean water you need"—and not to eat from the two cans of tuna they have left. Sougarret tells him that the rescue will be through a third, man-size borehole, most likely aimed at the workshop farther up the mine. Urzúa is surprised: He'd expected to be supplied and kept alive through the two new shafts while rescuers tried to open a path through one of the chimneys. "I never would have thought they'd get us out lifting us up through a hole."

Urzúa reports on these conversations to the other miners. A few are furious. "You can't tell them we're fine," they say. We are not fine. We're hungry, we're tired, we want to get out of this infernal place, the men say. If you lead those government guys in charge of the rescue to believe we're "fine" they're going to take forever to get us out of here.

At about 6:00 p.m. the second drill breaks through, just 1.3 meters (4 feet) from the first. (Three days later, on August 26, a third drill hole reaches the interior of the mine, up at the workshop at Level 135; it will play a critical role in the rescue.) The second borehole will become a permanent "utility" tube where electrical and fiber-optic lines will be lowered; and the first borehole will be used to lower supplies in the plastic PVC tubes that come to be called *palomas*, or pigeons. Bottles of clean water arrive, medicines, more glucose gel to drink. To monitor the supply hole and work unloading the supplies (and to keep everyone down below busy) the miners agree to be divided into three eight-hour work shifts. One, composed mostly of the contract mechanics, chooses Raúl Bustos, the meticulous tsunami survivor, to be their leader; the second and third groups, made up of men who sleep in and near the Refuge, choose the twenty-seven-year-old miner Carlos Barrios and the former soccer star Franklin Lobos. The trapped men of the A shift have a new energy, and sense of purpose. After eighteen days of crisis, Luis Urzúa decides to start acting like a boss once more, symbolically putting back on his white helmet.

On August 23, together with a *paloma* shipment of toothpaste and toothbrushes, the miners receive their first letters from their families. Many hear the same message Mario Gómez receives, news that might seem odd in the context of being near death, but which helps keep the men calm: The bills are all being paid, we haven't fallen behind on the rent or anything else, don't worry. Jorge Galleguillos reads words of support from a son who'd been estranged from him; Edison Peña gets a marriage proposal from his girlfriend; Carlos Mamani hears that he's a millionaire. Víctor Segovia hears from the daughters he's been writing to in his journal for eighteen days. "I had to keep taking pauses while reading," he writes in his journal afterward.

On the afternoon of the following day, August 24, a phone line is lowered back into the hole once more. Stand by, the voice says. We're connecting you to La Moneda, the presidential palace in Santiago.

President Sebastián Piñera has returned to his office in Chile's capital, and his long-distance call is patched into Pedro Gallo's makeshift phone, and thence to Level 94 of the San José Mine. He speaks to Luis Urzúa, and through him he reassures all thirty-three men that the government is doing everything in its power to get them out. Among other things, the government is accepting help from many other countries around the world. The Spanish prime minister and President Obama have expressed their support, Piñera says. Remembering what the trapped men had told him earlier, Urzúa thanks the president for his efforts, but then very quickly asks when the rescuers are going to be able to get them out of "this hell." *Este infierno.*

You won't be out for September 18, the president answers, and that's an immediate downer for the thirty-three men, because Independence Day is the biggest secular family gathering on the Chilean calendar. It's a bit like the Fourth of July and Thanksgiving wrapped into one, and this one was going to be an especially festive September 18, since it's also Chile's bicentennial.

But God willing, the president adds, we'll have you out by Christmas.

Urzúa makes a joke with the president about having the rescuers send them down a bottle of wine to celebrate the bicentennial, but after the conversation is over and the phone line disappears up the shaft again, several of the men slip into a deep depression.

"They thought we were going to get out right away," Urzúa says later. "Instead, we were going to be trapped maybe four more months." Urzúa studies the faces inside that rocky cavern. Jimmy Sánchez, the youngest of the miners, the teenager who's too young to work in a mine legally, appears especially stricken. Many of the men have regained just enough strength to rise to their feet, and the news of the wait to come returns grim and exhausted expressions to their sooty faces. The mine is still trembling and thundering around them, and at any moment a new collapse could destroy the life-giving 4.5- and 6-inch shafts that link them to the surface. A wait of four months in this oppressive heat and humidity might kill one, or two or three, of the weaker men.

"Four months!" Several of the men begin to shout at Urzúa. We can't wait here until December, there's no way, they say. We need to find our own way out of here. Through the Pit, someone says. As soon as we're strong enough we can climb out. The air of incipient mutiny once again fills the space, until Mario Sepúlveda begins to speak.

"You think I don't want to get out of here, too?" he says. "If I could, I'd grab on to the next thing that comes down that hole and pull myself through it. But I can't, because I'm too big." Perri speaks these words in the raspy, half-crazed, and ironic shout his companions know too well. It's the voice of a possessed lover of life trapped in a dungeon, a man who can joke about death, about eating other men, and about squeezing into a six-inch borehole. No, the only alternative is to wait, Mario says, and soon the other strong, calm men among the trapped thirty-three are repeating his words. "*Tranquilos, niños.*" It's what you say to two guys who are about to get into a fight in a bar. *Tranquilo.* We have to stay patient and organized, says Juan Carlos Aguilar, who's been saying that since August 5. We have to thank the Lord for the miracles that have already unfolded before our eyes, says José Henriquez. And we have to be prepared to live together in this hole until December if we have to, says Mario Sepúlveda.

Not long afterward, Luis Urzúa has his first discussion, via telephone and privately, with the psychologist, Iturra. "What comes now," the psychologist tells him, "is going to be the hardest part."

12

ASTRONAUTS

In the first days after contact with the men trapped inside the San José Mine, the professionalism of the Chilean doctors on the surface saves their lives. The minister of health, Jaime Mañalich, has assembled a medical team that makes a critical early decision: They will resist the entirely understandable desire (expressed by assorted drillers and local officials) to start "shoving food down the hole" (as one NASA doctor puts it) to the starving men below who are asking for it. A man who's been denied food for more than five to seven days is severely depleted in phosphates and potassium, which the body uses to digest carbohydrates. In the absence of these compounds, a hearty meal can trigger cardiac failure. This was the lesson learned in the final days of World War II, when GIs inadvertently killed many starving concentration camp survivors by feeding them C rations and chocolate bars. The Chilean medical authorities have consulted with experts at NASA and other agencies from around the world, and they will now heed their advice to go "low and slow," and begin to feed the men a mere 500 calories a day for the first few days, largely with an energy drink that's supplemented with potassium, phosphates, and thiamine, a B vitamin that the body uses up during starvation. Without thiamine, feeding the men could trigger Wernicke-Korsakoff syndrome, a nerve disorder that causes a catastrophic loss of muscle coordination. The Chileans send urine test strips down to the men, similar to the ones used by NASA to monitor the health of its astronauts. The strips test for "specific gravity" (an indicator of dehydration), ketones (an indicator of starvation), and myoglobin (which is produced when muscles break down). Sixteen of the thirty-three men test positive for high myoglobin: The eating away of their muscle tissue has triggered the early stages of kidney failure. Those men are sent down extra freshwater, and also cots to keep them off the hard surface of the mine when they sleep, because sleeping on rock is also causing their muscles to break down. (The Chilean government puts out a public call for portable cots that can be fit into a *paloma* tube and assembled below, and a local firm provides them.) Soon the men who were in danger of kidney failure are recovering. The Chileans handled the initial stages of treatment "in textbook fashion," the NASA physician James D. Polk will say afterward, "and because they did . . . out of the thirty-three miners they had not a single complication."

To further assess the state of the men's health, the doctors send down a scale. It's a harness-and-wire contraption, and the men attach it to one of the cherry-picker baskets they used when fortifying passageways. They raise the basket, sit in the harness, and hang in the air, suspended, another miner weighing each man as if he were an enormous specimen of pale-skinned produce. The smallish Alex Vega discovers that he's lost 16 kilos (35 pounds) and now weighs just 46 kilos (101 pounds), and the much taller Franklin Lobos is shocked to see he's lost 18 kilos (40 pounds) of the 86 kilos (189 pounds) he weighed before entering the mine.

The doctors ask Urzúa if anyone down below has experience in giving injections and taking blood pressure. The shift supervisor discusses this with the other men and someone remembers hearing Yonni Barrios talking about doing something like that.

"We need you to give some injections," Urzúa tells Yonni, and at first Yonni declines. "He's kind of pigheaded, but eventually we told him that had to be his job." Yonni talks by phone to the medical team on the surface, and tells them he's administered only one injection in his entire life: to his mother, who was a nurse, when he was fourteen years old. But he knows how to take blood pressure because Susana has high blood pressure and he's taken hers. Perfect, they say. You'll be our nurse down there, and very soon the Latin American media has dubbed

Yonni "El Doctor House," after a U.S. television program that's popular in the region. NASA has informed the Chilean medical team that living in isolated and stressful conditions, away from sunlight—trapped in a mine, or working in a space station—can cause both vitamin D deficiency and something called "latent virus reactivation." Yonni must therefore give his fellow miners vitamins and vaccinations against pneumonia, tetanus, and diphtheria. He performs this task with the calm gentleness that the women in his life have always admired.

The same tube that brings lifesaving vaccinations brings more private letters from the surface, and sends private letters back to the top. Víctor Segovia writes a message that is tinged with despair. "Panchito, I'm not going to lie to you about how things are down here. We're in really bad shape. There's so much water. The mountain is thundering a lot. This hell is killing me. I try to be strong, but when I sleep I dream that I'm in a barbecue, and when I wake up I'm in this eternal darkness. Each day wears you down." When his relatives read this, they decide to inform the psychologist.

In the diary he's keeping below, however, Víctor records some good news that other miners have received in their own letters. "One of our companions was told that Leonardo Farkas put 5 million [pesos] in a bank account in our names." Farkas, he writes, is trying to raise enough donations to make them all rich, "so that we will never have to work again."

Two days after the miners beg the president to get them out of "this hell," the rescue team sends a camera down to the trapped men so they can send back some video of what their hell looks like. Florencio Avalos operates the camera, and a bare-chested Mario Sepúlveda acts as tour guide, with a thickly bearded Alex Vega (in a very dirty and timeworn Dallas Cowboys jersey) as his assistant. The men shoot about thirty minutes of video, of which about eight minutes will be broadcast to a captivated prime-time audience in Chile that night.

The portion of the video that's broadcast begins with Mario showing Luis Urzúa sitting in his white pickup truck, making a schematic drawing for the rescue team above, and moves on to the Refuge, where two exhausted-looking miners, Osman Araya and Renán Avalos, are sitting atop the box that was once looted and that contained their provisions. "Here we have two very important miners, who are monitoring the *palomas*," Mario says in the cheerful voice of a talk-show host, as if he were trying to get the two men to ease their grim expressions. They stand up and open the box to reveal its contents, five bottles of fresh water recently sent from the top. "One way or another," Mario says, "we're trying to make a good organization out of this so that all turns out well."

The video cuts to Jorge Galleguillos, who lifts himself and sits up straight as Mario wakes him. His look is distant, dazed. Another man nearby remains asleep, his mouth open to the ceiling, even as the camera's light shines on him. Claudio Yáñez sits up and manages to smile and send a *saludo* to his family. Next the camera cuts to the Refuge, which Mario describes as "our cafeteria." Edison Peña looks directly into the camera and says: "Get us out of here quickly, please." Mario shows the audience a table where five men are gathered, a game of dominoes in progress. "This is where we have meetings every day, where we plan things," he says. "This is where we pray. Where we have an assembly to make decisions among the thirty-three of us." Mario approaches Víctor Zamora, the former orphan from Arica, the town on the border with Peru. "We don't know if he's Chilean or Peruvian," Mario jokes, and everyone laughs and Zamora breaks into the smile of an oversize child. Zamora led the attack on the food the first night they were trapped, but of course no one watching this video on the surface will know that. He addresses the camera, and looks and sounds more composed than any of the other trapped miners as he tells his family, with a reassuring nod, to "take care of yourselves" and that "all of us are going to get out of here." Then he addresses the rescuers: "We want to thank you all for having the courage not to leave us here, helpless." Zamora speaks like a man deeply at peace with himself and his difficult situation, as if he were some philosopher or motivational speaker who's wandered into the mine. "We heard what you did out there," he says to the miners' unseen rescuers. "And you know what, *niños*? We're going to give you all a big round of applause." All the miners around him start clapping, and the edited video cuts to a collective chant of "Chi-chi-chi, le-le-le," followed by Osman Araya praising God, and a rendition of the national anthem.

The video ends with Mario Sepúlveda, standing at the center of the frame, still shirtless, making an impassioned final statement. "This family of miners, my friends, isn't that same family you knew one hundred or one hundred and fifty years ago," he says. "Today the miner is an educated miner. He's a man that you can sit down and talk to. He's a man who'll stick out his chest, who can sit and talk at any table in Chile. A big kiss to all Chileans."

The video sets off a wave of national pride in Chile. Above all, people are moved by the sight of the half-starved and soot-covered Mario Sepúlveda, a Latin American everyman speaking with a kind of manic eloquence, and with great courage despite being trapped in what looks like a dank, dark, and foul place. In the weeks to come Mario will be referred to in newspaper and website articles in Chile and around the world as "Super Mario." His energetic video performance is broadcast far beyond Chile, too, making him the hopeful and unlikely figure at the center of a drama that fleetingly unites the globe, his raspy-voiced optimism offering a reaffirmation of the human spirit. Those are real men down there, 2,200 feet underground, and not mythical figures, even though there's something mythically epic about their story. They were given up for dead, trapped in stone, but this new video offers proof of how alive they are, how real and how dirty, looking desperate but sounding hopeful, stuck in a place of darkness and splashing water. These haunting images are broadcast and rebroadcast, looping their way across time zones and continents, as the tale of the thirty-three men is told and repeated by professional tellers and retellers of melodrama. When the world's televisions and computer screens are turned off, the story continues to circulate, in family and workplace conversations. *Did you hear about those guys, those miners in South America? Did you see them?* There are no instruments that monitor the collective subconscious, no global psychic seismographs, no gauges to measure the flow of humanity's dreams. But if such instruments existed they would pick up, in these final days of August, a substantial increase in the volume of nightmares and dreams set in caverns, graves, tunnels, and other dark and forbidding places.

Among those who know the thirty-three trapped men well, the effect of those eight minutes of video is markedly different. Jessica Chilla, who received that long, heartfelt embrace from her life partner and the father of their daughter on the day he went to work, sees the video and slips into an anxious state of near mourning. The Darío Segovia she sees on the video is not the man she knew. He's suffering a constant torment she can see in his wan eyes, in the way he holds his temples and turns away from the camera. His sister María, the "mayor" of Camp Esperanza, sees him and thinks that now more than ever he needs someone to embrace him. Most of the other miners don't look anything like themselves. To those who know him, Osman Araya is a man self-assured in his Evangelical identity; on the video he looks meek and wounded, and when he gives his speech about God, he is clearly fighting back tears. Pablo "the Cat" Rojas can be seen in profile, sitting on the ground shirtless, and he looks washed out and shrunken, as if he'd taken his square, middle-aged head and slipped it on top of the body of a boy. Jorge Galleguillos, known to his family as a man who's tall, stubborn, robust, and proud, can barely utter a word and is almost unrecognizable under the layers of soot and the fungi covering his skin.

The relatives of the thirty-three trapped men, and Chile itself, would be more disturbed if they were allowed to see what the government has edited out of the video—one news report will hint at what the rescuers know about the condition of the men when it says that five are suffering from depression and didn't want to appear in the video. In the unseen footage, Mario Sepúlveda sloshes though the mud, shows the viewers a decrepit bathroom, and begins to lose his composure toward the end as he summarizes their condition and their mood. "We're going to get out of here. We aren't going to stay here. Our families need us," he says. "We are very grateful, *chiquillos* . . . The only thing that we ask, personally—don't show the humidity and the conditions we're living in." He makes an allusion to death, revealing how close it feels to them. "This is a situation for warriors. If we have to give our lives for Chile, we'll do it here, or anywhere else . . . We'd be truly grateful if you could tell our families that we love them. There are a lot of beautiful people behind us." He then mentions his hometown, Parral, his neighborhood in Santiago, the sports clubs he belongs to, and says: "The people who know me know that I have a heart that's this big," opening his hands to form a circle before his bare chest. "And I'm going to take this little heart, and lift it to the top for the people in need. I'm going to keep on fighting, until the bitter end."

Finally, like many of the other men, Mario feels the need to speak directly to his family. "Francisco," he begins, though he can barely utter his son's name without beginning to cry. He coughs and continues. "My slogan: Dog. Braveheart, *huevón*. Mel Gibson, *huevón*. Papá will always be there to protect you, *viejo*. I swear to you that I'll always be there." With that, Super Mario is overcome with emotion and turns away from the camera, gesturing for Florencio Avalos to stop recording. When the minister of mining and the other leaders of the rescue effort see this video, they agree to respect Sepúlveda's wishes: After extracting a few scenes for public consumption, the officials store the complete video, and its most disturbing and sad images, in a government archive.

The psychologist, Iturra, examines the images in this first, extended video from below and finds reasons to be optimistic about the mental state of the thirty-three men. After some additional telephone consultations, and after assessing the first communications back and forth from the mine to the surface, he pronounces the group and its members in good shape. "*Están cuerdos*," he says. They're sane. A few members of the rescue team think the doctor himself is crazy for saying this, but the clinical evidence is pretty clear. "Psychologically speaking, they were all healthy," he later says. "They were scared, yes. To be scared under those circumstances is normal. But they weren't screaming to get out either." Among other things, the psychologist takes heart in the concern the miners have expressed for others—he heard it in that first conversation with the minister, when they asked if the truck driver Villegas had made it out; and in Mario Sepúlveda's and Víctor Zamora's speeches to the rescuers. The men haven't yet succumbed to panic and have retained some semblance of organization, and the psychologist is pleasantly surprised by this development, since studies of large groups of men and women confined in small spaces for long periods of time have often ended with people turning violent and walls literally covered with blood. Iturra, whose training is based on the client-centered philosophy of the American psychologist Carl Rogers, believes he can treat the men below as he would any client. "This is a collaboration and we're going to work together until you get out," Iturra tells the miners in phone conversations. "I'm here with you until the end." Whatever happened between the men in the seventeen days before they were found doesn't interest him, he says. "I'm not here to judge. You did whatever you had to do."

Iturra has the numerous case files on the thirty-three men that have been assembled by the Chilean medical authorities and social service agencies. In these records he can see many prior battles with the unique torments of mining life, and also the family and amorous puzzles that are common among Chilean workingmen. One of the men has a previous suicide attempt, two are epileptics, another was once diagnosed as bipolar—and from his own observations on the surface, he knows several have mistresses whose existence has been either confirmed or revealed to their wives for the first time in the improvised little village of Camp Esperanza. Iturra is a psychologist specializing in the mining industry and none of these things faze him much, because he knows that in addition to the stresses and sorrows of a miner's life, there is the fortitude, brotherhood, and sense of self-worth mining's hypermasculine culture gives him. But neither Iturra nor anyone else in Chile has experience in treating men suffering the extended isolation these men are going to have to endure. If they are indeed stuck until Christmas, they will have been trapped underground twice as long as any human being in history. They are like men on a mission inside a stone space station, or castaways on a lifeless planet, and to learn how men can endure such confinement and isolation, Iturra has consulted via e-mail with NASA. Albert W. Holland, a psychologist for the space agency, will soon arrive on a flight from Houston (along with two NASA medical doctors and an engineer). Holland has told Iturra via e-mail that he will have to prepare the men and their families for the long haul. "Long-duration thinking," he calls it. "We're looking at a marathon here," he tells Iturra, and soon the marathon metaphor is being disseminated by the Chilean psychologists to the men and their families. *Un maratón.*

On the International Space Station the astronauts have weekly video-conferences with their families, and the Chilean recue team prepares something similar for the trapped thirty-three men. For the moment there is no video link between the surface and the mine, so instead the rescue team asks each family to record a short video message to send down below. The psychologists tell them to communicate positive thoughts, not to speak of family complications, and these instructions are clearly on the minds of five members of Alex Vega's family as they gather underneath a tarp on the mountain to record their greeting.

"Hello, my love," begins Alex's wife, Jessica. This will be the first time Alex sees her since she denied him that kiss on the morning he went to work, and she speaks in a gentle and natural voice that belies all she's been through over the last three weeks: from the first days when she worked hard to keep her children on their school routine as if nothing had happened; to the worries of those final days of waiting when so many gave up the miners for dead. "Here I am, sending you all my strength, my *cariño*. Your children are well. They send you kisses." She says, in a mildly suggestive voice, that they'll have "a proper celebration" when he gets out. She is sitting before a Chilean flag that's decorated in the center with a portrait of Alex looking clean-shaven and movie-star handsome. Next to speak is Roberto Ramirez, the mariachi who wrote a song in Alex's honor. Roberto is the boyfriend of Alex's sister, and he addresses Alex by his nickname, Duck. "Crazy Pato. You gave us the scare of our lives. But when you get out, I challenge you to drink a tequila with me." His sister Priscilla says, in an upbeat voice: "Little brother, this is a lesson God is giving us. I hope you can see it. God doesn't send us a test we can't endure. And this one will be no different." She jokes about the beard he has in the video, which makes him look like "a wolf man." Alex's father, José, who entered the mine to try to get his son out, is wearing a white miner's helmet. "Son, I want to say hello to you, and all your companions, your brothers, because they're not just coworkers anymore, but brothers in this great odyssey that you all are living." More than forty of Alex's relatives have been gathered at the camp, José says, but couldn't be squeezed into the video. Then the elder Vega apologizes for not having more to say because "we Vegas are people of few words." A cousin adds: "Alex, you united the Vega family like never before," and everyone before the camera nods at this truth. Finally, they launch into the song they sang on the night before the miners were found, with Roberto the mariachi pointing to Alex's father, to his family, and to Alex's portrait on the flag to illustrate that the song is about Alex and all of them, and when the song ends Priscilla raises her fist playfully and everyone shouts: "And El Pato will return!"

On August 28, inside the mine, the thirty-three men huddle around the small screen on a video camera to watch these videos, their first glimpse of the outside world in twenty-three days. Luis Urzúa sees his wife, Carmen, and she looks exhausted and depressed, and a short while later he writes her a letter telling her to cheer up. For three of the men, however, there are no video messages—or so it seems. "They were very upset, especially José Ojeda, whose message was on the camera but they made a mistake when showing it," Víctor Segovia writes in his diary. "He got really upset and didn't even want to see it when they tried showing it to him."

The video the Galleguillos family sends down contains a surprise: Jorge's twenty-six-year-old estranged son, Miguel Angel, is on the screen.

Before the accident, "I had a problem with him. He rebelled against me," Jorge says. "We didn't have a lot of contact." Fatherhood is a challenge for any man, but what it means to be a miner and a father in Chile has changed dramatically in Jorge's lifetime. Jorge grew up watching his own father enter a small shaft mine alone in the 1950s, and he can remember being six years old, playing outside the entrance to a mine while his father worked inside. His father toiled underground with a pickax, and outside in the sun Jorge dug small holes and covered them with wood planks, making play "mines." Jorge's first paid job was at twelve, loading pack animals with one-hundred-pound bags; later he (like Darío Segovia) also worked carrying loads of rock out of a mine in a harness made of wolf's leather. Decades of underground work toughened Jorge, and if you asked him why he was still working underground into his late fifties, he might say, "Because the mine is for the brave!" ¡La mina es para los valientes! He worked hard to spare his own children the dangers and rigors of underground work, and his reward was a son who didn't understand why his father could be gruff and angry. On August 4, the day before he entered the doomed San José Mine, Jorge had called Miguel Angel, who was celebrating the recent birth of a baby boy. Years earlier, the first Galleguillos grandson had died not long after being born, but this new baby boy, or guagua, was healthy.

"I asked Miguel Angel about the guagua, and he said, 'What do you care?'" ¿Qué te importa vos?

"How is your little girl?" the elder Galleguillos asked.

"Why?" his son asked, irritably, and with that their final conversation ended. For seventeen days Jorge wondered if that would be the last time he talked to his son, the painful memory of Miguel Angel's brusqueness adding to the sense that he was leaving his life with so much unfinished and unsettled.

But the first letter down from his family had words of support from Miguel Angel. And now, on this video, Jorge Galleguillos sees sitting and standing before the camera his brothers, his sister-in-law, his niece from Vallenar—and his two sons, Jorge on one side, Miguel Angel on the other. The same Miguel Angel who was so curt with his father on August 4 speaks words of encouragement. "*Viejo*, take care of yourself," he says. "Just be as strong as you can."

After seeing this video, "I felt this amazing joy. This emotion," Jorge says. "But a little after that I started to get sick. I got depressed. It's hard to describe." Seeing one of the things he most wanted in life, an acknowledgment from his son, sends Jorge into a long funk. He struggles to explain why. Perhaps it's the realization that he had to be buried alive for his son to forgive him whatever he's done to make him angry. Or maybe it's the simple fact that he can't be with the son who is speaking with him again, or see his new grandson, or the rest of the family he now knows is reunited and gathered up on the surface. "I had this longing to be with them," he says.

Jorge has had to be brave and strong in his long mining life. Brave when he was twelve years old and entered a mine for the first time, and strong when he was an old man and the mine was wearing down his muscles and lungs. But he's never had to be brave and strong in this way before, with his body weakened by two weeks of hunger and humidity, dependent on others to feed him, the object of a family love that he can see but can't return. Suddenly the things that should make him feel good make him hurt. A few days ago he wanted desperately to eat something, but that chocolatey nutrition supplement the rescuers have sent from the surface to feed them, Ensure, is roiling away at his insides. "That milk took its revenge on all of us," he says.

When the Pastor, José Henríquez, drinks his bottle of Ensure he gets so sick he nearly passes out. Pedro Cortez sees this and gives his bottle away to another miner. Many of the men begin to suffer the first in a new series of intestinal and urinary disorders that will torture them in the days to come. Several of the men need to urinate, but can't, and eventually their collective suffering is bad enough that Mario Sepúlveda tells Yonni Barrios, their nurse, to demand a remedy the next time he talks to the doctors on the surface. Jorge Galleguillos has even less strength to deal with these complications. His legs are swollen and even walking a few feet is painful. A fungus is spreading over his body. At the daily meeting, he often can't rise to his feet, and the men stand over him and say a prayer for his recovery.

Víctor Segovia's diary has not always been upbeat in the days since the drill broke through and found the trapped thirty-three men. "Claudio [Yáñez] ruins my mood by sleeping all day and only wakes up to criticize . . . [Darío] Segovia almost got into a fistfight with Franklin," he writes on August 24. "Everyone's spirits are very bad. Before help arrived there was peace, we prayed every day . . . Now that help has arrived, instead of being more united, all we do is fight and argue . . ." Every other day, Víctor records new rumbling from the mountain, a reminder of the rockfall that trapped them. Escape from this aural torment is tantalizingly close, but for the moment all he can do is wait to be rescued—and to be fed. "Now I know how an animal in captivity feels, always depending on a human hand to feed it," he writes. Almost every day, his diary records another petty argument between the men, but on August 28, after seeing those first family videos, his mood is upbeat. "Everything is well organized . . . Our spirits are quite high today. We are very happy." After weeks in which their clothes have been soaked in sweat, and nearly all of them shed their shirts, a shipment of new nylon shirts arrives. They are the same red as the jersey of the Chilean soccer team, and many of the men begin to wear them.

On the evening of August 28, this newly uniformed team of Chilean heroes gathers for a meeting. "We spoke about a very private matter for when we get out," Víctor writes in his diary. "We're the only ones who know what we've lived through. We'll share it in good time." The subject of this meeting is the very story they're living. They've gone from being nobodies risking their lives in a piece-of-shit mine to talking with the president and his ministers, and even with a beloved Chilean icon, the soccer star Iván Zamorano, who had a brief phone tête-à-tête with Franklin Lobos (the two men were teammates for two years in the 1980s). The sense of their own celebrity is infectious, especially to Mario Sepúlveda, who has been making noises about how there's going to be a lot of money to be made in the telling of their story. Newspapers are starting to come down the shaft, and the men have seen a story comparing them with the Uruguayan rugby team that was stranded in the Andes; the article described how the Uruguayans sold the rights to a movie and book. Juan Illanes says the story of the San José Mine belongs to all of them, and they have to share in it equally, and this is so obviously true that no one argues with him. Illanes says further that they should keep to their previously agreed pact of silence about the accident and its aftermath, and that the diary Víctor Segovia has been keeping is a record of the struggle and that it should belong to all of them. All the miners agree: Víctor is their official chronicler.

The next day Víctor is writing with the new pen and notebook he's asked his family to send him. Mario Sepúlveda comes to speak to him. Víctor's diary is a "holy object," Mario says, it might be the book that tells their story one day, and thus worth money to them all. Víctor thinks about this a bit and writes: "I wrote this diary to survive, not to turn it into a book . . . I didn't realize it was such a big deal." Víctor, whose education ended when he got kicked out of school in the fifth grade for fighting, never imagined that putting words to paper could make him feel better about himself or that he would one day think of himself as a writer. He's never traveled beyond the desert cities and towns around Copiapó, but underground he's become the chronicler of a story that will one day circle the world.

In the trailers and bungalows on the surface of the San José Mine where the leaders of the rescue have set up shop, it seems as if the entire world is moving to support them. "We could ask for anything, from anywhere, and people would try and get it to us," Cristián Barra says. Drilling experts and drilling technologies from around the world are headed for the mine: from Johannesburg, South Africa, and Berlin, Pennsylvania;

from Denver, Colorado, and Calgary, Alberta; and from a U.S. Army forward-operating base in Afghanistan. The Chileans are planning on restaging one of the great rescues in mining history—at Quecreek Mine in Pennsylvania in 2002—and they're going to need a global cast to pull it off.

At that American coal mine, nine men were trapped when they inadvertently triggered an underground flood. They were rescued by the drilling of a vertical shaft 240 feet deep and thirty inches wide, through which they were raised up in a steel basket. The Chileans will need to drill a hole that's eight times as deep. To do so, they summon one of the biggest drill rigs in Chile, an Australian Strata 950 raise borer, from the Andina Division of Codelco in the nearby Fifth Region of Chile. Unlike the smaller machines that have drilled three diagonal boreholes to reach the men, this 31-ton machine can only drill straight down, and it begins working on August 30, near the spot once judged too unstable when Eduardo Hurtado's Terraservice team tried to drill its first hole three weeks earlier. This is the drill that will take until December to reach the men. But even before it begins working, Igor Proestakis, a mechanic with one of the drill teams, presents an alternative plan to his bosses. Why not take that third, six-inch drill hole that reached the men at the workshop, and simply widen it, using a series of ever-bigger drill bits? A few days later, André Sougarret approves the drilling of this second rescue hole, to be called Plan B. It will be drilled in two stages, the first twelve inches wide, the second twenty-nine inches wide.

There's one problem with Plan B, however: No one in Chile has ever drilled a twenty-nine-inch hole sideways that's as deep as the one they'll need. The best machine to drill such a large diagonal shaft is a Schramm T130XD, and soon one such rig is on its way to Copiapó from the Collahuasi mine, more than 600 miles to the north. After soliciting advice from mining companies around the world, Sougarret's team decides that to speed up the drilling it will accept an offer from Center Rock Inc. in Pennsylvania to use one of its clustered drills, which is essentially four drill bits, each the size of a volleyball, attached to a single shaft. This equipment will be used to drill the second phase of Plan B, and it weighs 26,000 pounds. UPS agrees to ship it for free: first overnight by truck from Pennsylvania to Miami, thence on a flight to Santiago, and finally by truck to Copiapó, where it arrives on September 11. There are very good drillers in Chile, of course, but the depth and angle of the borehole needed will require an operator who can make the equipment do things it wasn't designed to do, so the Chileans reach out to the Kansas-based drilling experts at the Layne Christensen company. Who is the best driller you have on a Schramm T130? The company calls Jeff Hart, who's in Afghanistan, drilling holes to provide drinking water and showers for U.S. troops.

Plan A and Plan B will soon be in motion, but both will push their drilling equipment beyond design limits and the probability of failure is high. Sougarret decides, then, to add a Plan C: a massive oil-drilling rig that will take days to assemble, but which will actually drill faster than Plan A once it gets going. The Canadian company Precision Drilling Corporation has one such machine sitting idle in southern Chile; thirty-seven truckloads are required to bring Rig 421 from a work site a thousand kilometers away. The technology that will guide it to its target and some of the technicians will come from the South African company Murray & Roberts.

In a series of phone conversations, Sougarret outlines the three drilling plans to Luis Urzúa. It doesn't really matter which one reaches them first, of course. But when it's all explained to Urzúa, it seems the Plan B drill will likely be the fastest.

The government of Chile is marshaling resources from dozens of different agencies, and several top-level officials are living at or near the mine site, including two members of the president's cabinet. The entire operation has the feel of a Chilean moon shot, and like an expedition to outer space it should have a name. The minister of health, Jaime Mañalich, telephones President Piñera and suggests they call it "Operación San Lorenzo," after the patron saint of mining.

The president does not warm to the idea, according to Carlos Vergara Ehrenberg's behind-the-scenes account. "Lorenzo," the president repeats. No, that sounds too much like Laurence, as in Laurence Golborne, he says. At La Moneda, the president's advisers have continued to conduct private polling on the rescue, and recently the numbers have begun to show something of concern: As high as the president's approval rating is, Minister Golborne's is higher. Golborne, who once occupied one of the more obscure posts in Piñera's cabinet, has become the face of the rescue, and he's starting to outshine his boss.

"Let's call it 'Operation Jonas the Prophet,'" the president says. The name never catches on, however. Instead, the Chilean and world media assembling at the site continually refer to the operation to reach the buried men by the name that's already been adopted, informally, by the men and women working in the rescue crews: "Operation San Lorenzo."

To keep up the spirits of the men living below as Operation San Lorenzo unfolds, and to keep them from turning on one another, the psychologist Iturra has access to a powerful, joy-inducing medicine: the voices of the people they love, as heard in a telephone connection. One by one on August 29, wives, brothers, mothers, sisters, fathers, sons, and daughters are guided past the lines of police officers and the security fences, to a small metal communications shack, roughly six by eight feet wide, that rests on the sandy dirt just a few paces from the hole that reaches to the thirty-three men below. There is a camera and a microphone, linked to another camera and microphone in a gallery inside the mine. In preparation for this moment, the psychologist has written a letter to the trapped men. Among other things, he's told them that those with two households, or with amorous complications, should give preference in these teleconferences to their wives and families. "I told them this because I could already see the conflicts with the girlfriends that were unfolding in the camp," he later says. "And I also told them it would be easier that way: because, after all, lovers and girlfriends tend to be a lot more forgiving than wives are."

Some of the girlfriends are having trouble getting into the camp anyway. Susana Valenzuela, Yonni Barrios's girlfriend, says that not long after the Farkas money showed up, Yonni's wife, Marta, "betrayed me" and the Carabinero police escorted her away from the new Camp Esperanza that's been built to give family members privacy, away from the media. On August 28, the Associated Press photographs Susana outside the camp, holding a sign with large letters that read "The Courage to Be Present," along with a picture of Yonni and much smaller letters that say: "For you, my love, your Chanita." The caption that circulates around the world calls Susana Yonni's "wife" but just a few days later the truth will come out, as social workers attached to the rescue effort realize that Yonni is in fact married to another woman at Camp Esperanza, and then learn from Yonni himself down below that he hasn't been living with said wife (exclusively) for years. Marta is photographed in Camp Esperanza holding a poster covered with portraits of Yonni, and journalists will begin using their imaginations to deduce—and report as fact—that Marta first became aware of Susana when they met at Camp Esperanza after the accident. In truth, they've known about each other for quite a long time.

Susana is both determined and crafty, and to get into those parts of Camp Esperanza closed off to her, she resorts to a spy-like deception. She sees a load of fish and vegetables being delivered to the big kitchen that's making food for the families and rescue workers. "I put on an apron, and grabbed a fish and an onion, and I walked right past the guards," she says. "The journalists saw me inside and asked if I was a relative, and I said, 'No, I'm just a cook.'" In this way, Susana will eventually work her way into that shack to talk to Yonni, despite the psychologist's advice.

At any rate, the miners and their family members can't talk about much in these first conversations, since Iturra will limit their first call to about fifteen seconds; their next call will last about one minute. Iturra is thinking about the mental marathon the miners will have to run and he believes that, as with the miners' food, a smaller dose of family love at the start is better. "In fifteen to thirty seconds you can't convey information—there is only the personal encounter. You're present," he says. "You say 'I love you, you can count on me.' And that's it. You don't have time to say, 'Your father is sad, your grandmother is sick, your son isn't going to school.'"

Iturra is following advice from NASA, but the thirty-three men aren't astronauts, they didn't volunteer to be trapped in a hole for months. After the too-short phone calls, the miners begin to feel, understandably, that they're being treated like children. Let us talk to our wives, our kids, they say. We're men: We're not helpless. The paternalism of the psychologists is plain to see in a video, shot on the surface, in which one miner's wife is speaking to her husband on the phone in that small shack.

"*Hola*, my love," the young woman begins, in a weak voice.

Iturra is sitting nearby. The woman is about to be overcome with emotion, it seems, and he quickly snaps at her to cheer up. "¡Ánimo!"

"Everyone is fine," the woman continues, sounding a bit more upbeat. She lists all his relatives, then says "I miss you" in a voice of hinting at despair.

"¡Ánimo!" the psychologist orders, and the young woman again tries to sound more cheerful, until a few seconds later the psychologist says, "Start wrapping it up."

Even after the rescuers establish a permanent fiber-optic link with the surface—which will include a television feed and uninterrupted phone connection—the psychologist will continue to limit the contacts with the family to roughly eight to ten minutes a week, which is about the time NASA gives its astronauts. (Eventually, Víctor Zamora will lead a brief "strike" against the psychologist by turning his back to the camera and refusing to talk to his own family until the psychologist grants all the men longer time for the video chats.) Contact with the outside world "takes you out of your reality," Iturra says. "It puts you in a world where you have no power." Iturra is trying to protect the miners from the feeling of helplessness: They can do things below to aid with their own rescue, but they can't be home to be good fathers or good sons. At home they are needed and they are rich, they are famous, and their kids need to be fed and protected. The men can't be in that world, but even with Iturra trying to shield them, they are pulled into it, because, despite the miners' many suspicions, no one is censoring or monitoring the letters that go down below. Via the "mail" that flows down the *palomas*, Zamora learns that his young son is being bullied in school: "Your dad is never going to get out! He got crushed under a rock!" Franklin Lobos learns that his ex-wife is up on the surface, and that his children are hoping he'll make amends with her. Others learn that the women in their lives have heard the voice of God and have decided they should take the next step and just get married already. In one of the first letters Edison Peña receives, his girlfriend, Angelica Alvarez, brings up the topic of marriage, to which Edison answers: "I don't understand why you'd want to marry me . . . I've had a lot of time to think about all the things I've destroyed, and about all you've suffered thanks to me . . . But I wouldn't want you to be with anyone else either and I'd like to make you happy, even though I've never managed to do that."

That letter finds its way, somehow, into the Madrid newspaper *El País*, and Edison's confessional is soon circulating around the Spanish-speaking world. The miners may not be entirely powerless to help their families—they can now send instructions and keep tabs on things, at least, via the phone—but they are undeniably without protection against the media. Some reporters are willing to pay their families for a glimpse at the miners' letters, but most simply charm them away, and soon the Chilean newspapers are reprinting many of the words written by the men below. Those letters then recirculate down into the mine, because despite the miners' suspicions, Iturra and the other leaders of the rescue team on the surface have decided that they shouldn't censor the newspapers either.

The Santiago newspapers are rolled up and stuffed into the *palomas* by workers at the regional governor's office who are in charge of sending the men reading material. Opening these precious relics of the surface world for the first time, the men can see just how famous they've become, the way their pictures are covering all the front pages. Yes, some of the more prudish surface workers have cut out assorted photographs and advertisements featuring scantily clad women, but no one stops the August 28 edition of *La Tercera*, for example. It has a feature article about a miner whose newfound fame is leaping off the page, and it quotes from a letter he's written from inside the mine. When thirty-two other miners read these words, they get an unexpected glimpse into the mind of Mario Sepúlveda.

13

ABSOLUTE LEADER

Despite its name, *La Tercera* is the second-most influential newspaper in Chile. Its August 28 edition carries a big spread on Mario Sepúlveda, reported and written in the hours after the miner's stellar appearance on Chilean and global television. The story says Mario's picture has been on the front page of *The New York Times*, *The Guardian* of London, and *El País* of Madrid. It quotes his speech from the August 26 video and interviews his wife, Elvira. "She is not surprised by the qualities of a natural leader her husband possesses," the writer says. The story quotes from a letter Mario has sent to his family in which he describes how the miners are getting along. "I am the absolute leader," it begins. "I organize things, give orders, and, as always, I avoid losing my temper. But the most beautiful thing is that I am respected and nothing is done without me knowing about it." Elvira says that a social worker from the governor's office stole the letter from her and gave it to the newspaper, but many of the miners' families doubt this. Stuffed into a *paloma* along with many other newspapers, this story reaches the men below and is quickly passed around. They read about themselves under the gray glow of artificial light, holding a page in which Mario is looking back at them from inside the very cave in which they are all trapped.

Fairly or not, to the trapped men the news story smacks of self-promotion. Mario was one of the first people to mention how rich they might become from telling their story, and this article suggests to some that he's trying to concentrate the media spotlight on himself, with his wife setting him up to be a media star when he reaches the surface. The men find his statements both amusing and insulting. Here they thought they were thirty-three men making decisions together, but the rest of the world is being led to believe that Mario is their "absolute leader." At this point, they've been stuck underground for nearly four weeks, each man struggling to keep his sanity, several trying to find a way out, all of them concerned for the welfare of the others. Yes, more than once Mario has stepped forward to do something that's helped save them, but always working with other men: When he climbed up the chimney to try to find a way out, Raúl Bustos was there with him; when he issued his angry call to prayer it was José Henríquez and Osman Araya who actually led the prayers. And for every time Mario spoke up and lifted someone's spirits with his pleading voice, there was another time when he broke down in tears and despair and his coworkers lifted *him* up. But in this story, in a newspaper that reaches every corner of Chile, Mario Sepúlveda is claiming to be their captain, their hero.

Several men, and especially the mechanics, see the letter and the newspaper story as evidence of Mario's manic need to be the center of everything, and they grow more suspicious of him than they are already. Raúl Bustos begins to mercilessly tease Mario about his boasting every chance he gets.

"Raúl Bustos started to call me out and make fun of me and laugh at me," Mario says. "He'd say, 'You're never going to be anyone's boss. Who do you think you are?' José Aguilar did, too."

Mario explains to his angry coworkers that he wrote that letter to keep up the spirits of his son, the boy he desperately needs to protect: He made himself into the one and only leader because he wanted Francisco to believe his father was his "Braveheart," his Mel Gibson leading men into battle. But Mario's explanations can't undo the damage to his underground reputation, and his letter sharpens the divisions among the thirty-three men.

Those who've slept in and near the Refuge continue to support the man with the heart of a dog. "The leader we had inside was Mario Sepúlveda," Omar Reygadas later says. "He kept us going. We can't deny that to anyone, and I'll never deny it, because I'm not an ingrate." Franklin

Lobos will listen to Bustos's digs at Mario and accuse Bustos of "deliberately dividing the group." Mario himself believes his enemies are working to "*mariconear*" him, a Chilean idiom that means to conspire against
someone, and which is derived from a slur for homosexuals. Never one
to sit by while others work against him, Mario decides to "put my cards
on the table" and marches up to Level 105 to confront them.

"Luis Urzúa was there, Juan Illanes, Jorge Galleguillos, all of them.
I went in, and I said, 'Look, you motherfuckers,* let me make this clear.
I am not the boss. But the boss, assholes, is that idiot who's worrying
twenty-four hours a day about these guys, about the guy whose belly is
hurting and needs help. The boss is the *huevón* who keeps everything
clean, and the boss is the idiot who has to tell the guys to clean their
work area. The boss is the *huevón* who just came from Level 120 and
put gloves on to clean the shit these guys left all over the place where we
all go to the bathroom, and because one idiot took his own shit and
covered the door with it. And do you know which *huevón* is the *huevón*
who does all that? It's me, you motherfuckers.'"

Later, Mario gets on the phone to the surface and chews out the
psychologist, who he blames (with no evidence) for the release of his
letter to the press. "Motherfucker," he begins. "What kind of professional
are you, asshole, to allow a letter to be passed on like that?"

Even as Mario tries to sort out the mess he's created, some take note
that he's monopolizing the phone link to the surface, and isn't subject to
the time limits that the other miners have. Even people who like Mario
believe that his sudden fame is going to his head. Víctor Segovia describes in his diary how Mario is pacing back and forth, frustrated,
because he's become a celebrity but he's still stuck in a hole and can't
do anything with his new fame. Among those who don't trust Mario,
it's Raúl Bustos who is most willing to speak out about his suspicions
and fears of the man with the heart of a dog. He believes Mario is a
common street fighter, the kind of guy whose brawling might have easily landed him in jail. Since the drill broke through, Bustos has been

---

*In Spanish, Mario said "*concha de su madre*." This means, literally, "your mother's shell,"
though its full, vulgar sense is more akin to "your mother's cunt." *Concha de su madre* is such a
common oath in South America, however, that I've rendered it here as the slightly less offensive
"motherfucker."

listening to Mario and Víctor Zamora make disturbingly violent jokes about the days in the very recent past when they were all starving to death. "They said they had a pocket knife and they were going to use it to slaughter people [*faenar*]. That they would have eaten certain people, or the first person to fall. They said it was a joke, but those are things you shouldn't joke about . . . I took the measure of them. I could see that they had this cruel streak." Bustos believes, rightly or wrongly, that the mechanics' sense of rectitude has kept the shift supervisor, Luis Urzúa, from being overwhelmed by Mario Sepúlveda and his "clan" in the Refuge. He's concerned about his personal safety, especially now that he's earned Mario's hostility, and he reveals this to his wife in his letters. "Raúl said he never slept well," Carola Bustos says. "Because he always slept with one eye open."

Several miners have spoken to the psychologist, Iturra, about the perceived bullying from other miners. "You can't even talk, because there's people controlling what you say," one tells him in one of the many individual phone sessions the psychologist has with the men. "I'm afraid."

"Get close to someone who can take care of you," the psychologist counsels.

The verbal jousting continues and every day Víctor Segovia details a new argument in his journal. One night, Claudio Yáñez gets in a loud disagreement with Franklin Lobos—Franklin has been "really moody," Víctor writes—and Claudio goes to bed with a pipe next to his cot because Franklin has threatened to hit him. "During the twenty days that we were starving and in despair we were always united," Segovia writes, "but as soon as the food started arriving and things got a little better, their claws came out and they want to prove who is tougher."

For the psychologist, it's obvious that the men are divided and that the fear among them is a natural product of the "crisis of authority" down below. He's learning about the conflicts from his phone conversations with the miners, and from his consultations with family members who've received troubling letters from the men. Urzúa is a "passive leader," and in the absence of a strong authority figure, "there were some people taking authority for themselves, and others doing whatever they wanted," the psychologist says. "Down there, if anyone got out of line," one of the miners will reveal to Iturra afterward, "a group of five or six of us would stare him down, and we would impose ourselves [*hacíamos fuerza*]." As some of the men try to sleep on the new cots provided by the rescuers, their thoughts are unsettled by this new fear: The idea that they're trapped in the mine not just with brothers in suffering but also with men who don't respect them, or who might attack them in their sleep, or who might betray the group and take the riches that await on the surface.

"I think it's because of fear that we're all bickering," Víctor Segovia writes in his diary on August 31. Víctor also believes that money waiting outside is causing some of the men to lose their heads, and he's grateful that his family never mentions money in their letters to him. On that same day, the topic of the arguments among the men comes up in the daily prayer at Level 90. "We prayed and asked that everyone keep their cool and that we stop arguing so much," Víctor writes in his diary. A few days later, thirty-three crucifixes arrive in a *paloma*. They've come from Rome and have been blessed, the men are told, by Pope Benedict himself. Víctor hangs one up on a box over his new inflatable bed and prays for peace among his brothers.

The thirty-three men are certainly not proud of the conflicts that have divided them in this, their fourth week of captivity. But it's hard to believe any other group of thirty-three people would have done much better under the circumstances. Imagine being sealed up in a hot and humid cave, subjected to about three weeks of deprivation and hunger, followed by a global media circus that you must endure while remaining confined in a mountain whose innards rumble routinely, suggesting that the whole story might just end with you dead and buried anyway. Imagine being famous and wealthier than you've ever been—but also dependent on strangers who decide what and when you eat and how long you can talk to your family. And imagine the pressure that comes with having an entire nation look upon you as a symbol of courage and all that's good and resilient about mining, a craft that's at the heart of your country's identity.

The men can see what their story means to the Chilean people in all the newspapers reaching them, and they feel the responsibility of what they've come to symbolize: endurance, faith, brotherhood. That's why, despite the many harsh words between them, most don't give up trying to be the proud and united Chilean workingmen the outside world believes them to be. In a certain sense, that's the way it always is in a mine, where being confined in a life-threatening situation with other men who insult and mock you is part of everyday work life. "In a mine, when you can treat someone poorly, and he's still there the next day, without holding resentments, when you sense he just wants to move on—all that generates trust," Iturra says. "You think: *This guy's not going to let go of me.*" As long as the men can keep busy, as long as they can still feel like miners, they should be able to keep at least a semblance of unity.

In fact, the men do fall into a work rhythm, one that's completely different from the routine in the mine before August 5. They unload supplies, medicines, and personal packages coming from the surface around the clock, and also maintain the communications link to the top and keep the lights going. Unloading the *palomas* brings all sorts of interesting things. Cowboy novels, pocket-size Bibles, and an MP3 player for one miner who has been complaining so much about everything, the other miners give it to him just so that he'll keep quiet. But then some of the other miners complain about the one miner with an MP3 player, and soon they all have one. For group entertainment, a Samsung SP-H03 Pico portable projector arrives. It fits in the palm of a hand, and the men will soon use it to watch videos, movies, and live television images projected onto a white sheet. But best of all, the *palomas* are starting to bring real food. The daily intake supplied to the men has increased from 500 to 1,000 calories now, and soon it will be 1,500. And the men are getting real meals, prepared by a kitchen up on the surface, including rice, meatballs, bread, chicken, pasta, potatoes, and pears, all in small but delicious portions.

After a few days in which the men heartily and thankfully devour this real food, the rescue team on the surface finds an uneaten pastry inside what should be an empty *paloma* sent back from the bottom. One of the men down below has returned that day's dessert. This thing you gave us, an attached note says, it isn't very good. Do you have anything else? The rejected dessert is an unequivocally good sign: The men are no longer so desperate they'll eat anything you give them.

•

On August 30, a day before the men pray for peace at Level 90, the rescue teams begin drilling the first hole designed to bring the men out. The Strata 950 raise borer is a machine so large and elaborate, many different metaphors are required to describe it. At nearly three stories tall, its general structure is that of a monument, or a gazebo, with six stainless-steel pillars, each about two stories tall, holding up a large white metal roof that itself has four additional white columns protruding from the top. This edifice rests upon a floor of recently poured and freshly set concrete, and it houses a series of hydraulic levers and shafts designed to guide man-size drill bits into the earth. The Strata 950 will begin by excavating a 15-inch pilot hole, and once that's completed, a second drill bit will widen it to 28 inches for the rescue cage. The first, smaller drill bit is composed of a series of interlocking beaded discs, and these begin to grind into the stone, creating a hole that is filled with 9.5 gallons of water per second to reduce friction. The Strata 950 crushes and sloshes its way downward into the diorite mountain as men in yellow overalls tend to it, working in teams to lift, turn, align, and lower a variety of heavy steel components, each man doing something different, as if they were working at a rock-crushing assembly line. The noise the machine makes, however, is similar in volume and pitch to that of a jet engine taxiing on a runway. The bit turns at the sedate but steady speed of about 20 rpm, working under the sun and then into the night, the site illuminated by white lamps that leave the work crews looking like extras in a sci-fi movie. The rescuers toil in a bubble of light, working to reach a group of no-longer-hungry but deeply irritated miner-astronauts waiting to be liberated, 2,100 feet below.

The NASA team of health experts and engineers arrives in Copiapó on the morning of September 1, after a two-day journey from Houston. On their drive from the city of Copiapó to the San José Mine, Dr. J. Michael Duncan takes in the dry and treeless landscape, and the colors and textures of geological features that look transplanted from Mars. He remembers that the Chileans are building a facility in this same desert, the Moon Mars Atacama Research Station, in which the harsh, waterless surroundings serve as a laboratory for studying the possibility of life on other planets. They enter the mine property and note immediately the frenetic activity at the site, the men and women in mining helmets and overalls. At the top of the hill, they see the massive Plan A drill at work. The NASA experts visit the site for several days, and on September 4 they are talking with the Chilean rescue officials in one of the small offices on the site when they hear loud cheering outside. They open the door and see people clapping and waving as a series of trucks pull through the gate: The first drill required for the two-stage Plan B has arrived.

The Chilean officials ask the NASA visitors to talk to the miners. Many of the assorted dignitaries who've come to visit the mine have found themselves guided to the communications shack for a moment to address the men down below: from the senator Isabel Allende to four members of the Uruguayan rugby team who survived a plane crash in the Andes. Albert Holland is handed a phone by a Chilean technician. "*Hola*," Holland says, and then stops, because he doesn't have much more Spanish than that, and he doesn't understand the rush of Spanish that comes in reply from down below. "Just say '*bien*,'" one of the Chileans on the surface says, and Holland says "*bien*" and soon the conversation is over. The NASA representatives meet with the families, too, and are introduced as members of the American space team that's come to lend their expertise to the rescue. Holland says that the rescuers are doing everything within their power to bring the men to the surface. A fire-plug woman of about fifty with sunburned skin steps forward; she's introduced to the NASA men as the "mayor" of Camp Esperanza. María Segovia has listened to Holland's speech and is moved, and the woman from Antofagasta who sells pastries at the beach gives the space expert from Houston a heartfelt embrace. "I'll adopt him right now," she says.

After sunset falls, the NASA men take in the spectacle of the night sky. This is a desert where astronomers have come for decades because they feel closer to space here than any other place on Earth. "The Milky Way stretched in a great arc from a set of silhouetted hills behind us all the way across the sky to the hills in front of us," Holland says later. "It was like standing under a brilliant bowl. The desert, the night, and the stars were completely silent . . . Brilliant, ageless, and still." Beneath the infinite canopy of the cosmos, Holland observes, there is "the feverish, human intensity" of the mine and its teams of rescuers working to lift thirty-three knuckleheads out of the mountain.

Down below, thirty-three trapped men who cannot yet cast their eyes on the Milky Way spend hours trapped in the stillness and the heat that comes from the center of the Earth. They cannot yet hear the Plan A drill, and the quiet is broken by the occasional moans of men who are not yet fully themselves in body and mind. They may no longer be hungry, but now that they're drinking clean water from the surface, and a lot of it, a few cannot pee. They are bloating up, and as they squeeze their full, aching, and stubborn bladders they direct their complaints to their medical volunteer, Yonni Barrios, who takes to the phone and talks to the surface. The medical team from the Ministry of Health listens to Yonni, and then asks him: Have you ever inserted a catheter before? The best treatment for urine retention, they tell him, is to take a tube, insert it in the patient's penis until you reach the bladder, and drain the contents. Yonni is sent down the catheter and gloves necessary to perform this operation. "If you tell me how to do this, I'll give it a try," he tells the doctors, even though it goes without saying that having to insert a tube into his coworkers' penises was not something he expected to find himself doing when he showed up for work on August 5. Fortunately, before he can perform this uncomfortable and embarrassing procedure, the doctors tell him: Wait, we'll send down some medicine first. Yonni, meanwhile, tries a home remedy: hot compresses, which he prepares by taking a few water bottles and heating them up by placing them in the exhaust pipe of one of the pickup trucks and starting the engine. "It was just hot enough to heat the bottle without melting the plastic," he explains. Yonni gives these hot water bottles to Víctor Segovia, whose suffering from urine retention is the worst, and helps him place the warm bottles between his overalls and his pelvis. After a few hours of this treatment, Víctor is able to produce a trickle of urine. Yonni reports this to the surface, and they ask him to send a sample to the top for analysis.

Next, Yonni puts on his gloves to treat the most serious, endemic medical problem the men face: the spread of fungus over their bodies. Before the rescuers reached them, only a few men suffered from this skin ailment, but now that they've been able to bathe, they've lost the layer of dirt and grit that was protecting them against the fungus that's literally raining down upon them. The continual flow of wastewater produced by the machines drilling on the surface, combined with the mine's routinely oppressive heat and humidity, has transformed the caverns in which they are living into a kind of fungus factory. The mud is beginning to rot and, when the breeze shifts in the mine, Yonni can smell the putrefaction. "It smelled like when you go to a river and pull up the black mud at the bottom." He can see fungi growing and spreading on the roof of the corridors and of the Refuge. "It was like these thin little hairs falling from the ceiling," he says. Hyphae, these filaments are called. "They would fall, as if it were raining. They were shiny, and they would glow when you put a light on them. They were like little transparent hairs." The fungus falls on the men when they're sleeping, shirtless against the heat, and when they're awake it covers their new inflatable cots and begins growing there, too. Angry red circles begin to cover their bodies. Yonni puts on gloves and studies how the fungus is invading the backs and arms and chests of his coworkers. Each sore is a few millimeters in diameter, with pus in the middle, and as time goes on they seem to penetrate deeper into the skin, despite Yonni's patient and constant treatment with creams. Yonni worries that the red pustules will soon grow infected, and he can imagine the fungus digging its way even deeper into the skin, spreading an infection that he'd be powerless to stop in this damp and fetid cavern. Yonni worries that if they are stuck here until December, this fungus could begin to devour them all from the inside and kill half of them off. They'd die and then be eaten by this living thing that thrives on moisture and darkness and the miners' own ever-paler flesh.

14

COWBOYS

Why not just leave right now? Why wait until Christmas for these government rescuers to come and get us? The question arises again and again in these first days of September. The thunder blasts and the wailing of the mountain have not stopped, and for many of the men the intermittent sound is fueling a deep psychological torment. They tremble in the dark as they try to sleep, and they don't feel like themselves, even now that they're fed. On August 5 the mountain sent explosions of rock and rolling boulders down the mine's passageways to kill them, and each new underground quake or tremor they feel is a reminder that it might try again to kill them. The unending aural and seismic torture inflicted by the massive mountain in which they are trapped is slowly eating away at their sense that they'll ever be truly happy and free men.

Yonni Barrios says they can find their way out through the chimneys or the cavern of the Pit. "Since Yonni thinks he's a badass [*se cree capo*] he's planning on escaping through the chimneys even though he knows they are blocked. And the ones that are not blocked have no ladders," Víctor Segovia notes in his diary on September 6. Yonni goes on about this so long, he convinces Mario Sepúlveda and older miners like Esteban Rojas that it might actually work. "Yonni is scared and desperate and is dragging innocent people with him," Segovia writes in his diary. "I think that all who follow him are headed straight to their deaths." Listening to the men talk with increasing detail about escaping leads Luis Urzúa to call to the top and request the telephone intervention of the one man on the surface all the miners trust completely: Pablo

Ramirez, the night-shift supervisor who tried to rescue them in the hours after the collapse. From the surface, Ramirez speaks to the group to "clear up any doubts" about the possibility of escape. All exits linked to the Ramp are completely blocked, he tells them, and up at Level 540 the mine is still collapsing, producing rockfalls that could kill either the men or the rescuers trying to reach them should the men get stuck in their ascent. With that, all talk of escape ends, for the moment.

Today, during their thirty-fourth day underground, as during their first night trapped, Luis Urzúa believes he has managed to keep the men under his supervision from killing themselves. But Víctor Segovia and many others are still frustrated with him. "We're baking down here, and always arguing, but when the people on the surface talk to our boss, he always tells them that everything is okay." Urzúa is an outsider, a shift supervisor with only a few months at the San José, and before the collapse he didn't even know most of the men, Segovia writes, and he's clearly siding with the mechanics (another group of outsiders, in the eyes of the northerners who've worked in the mine the longest). The mechanics, for their part, are so frustrated with the old-timers and young "clan" that sleeps in and near the Refuge that they've stopped recharging the batteries to their lamps. "Now we are without any light," Segovia writes.

The only thing Urzúa can do to mollify his anxious men is to pass on what's he's learned from Sougarret and the others about the rescue efforts. People and drills and equipment are coming from the United States, Austria, Italy, and, of course, from the biggest mines in Chile. Then, on September 7, Urzúa learns that the Plan B team has reached 123 meters in the first stage of its two-stage drilling plan, surpassing the depth of the Plan A drill after just two days. If the Plan B team can keep going at this pace, it's likely the rescue capsule will lift them out of their hell long before Christmas. This information helps calm the men considerably. Another potential revolt has been averted, but Urzúa has plenty more things to worry about it. The shift supervisor is constantly being called to the phone, to talk with all sorts of people who have little, if anything, to do with the rescue. One day it's the ambassador of Palestine in Chile on the line; another day, it's the ambassador of Israel. Urzúa talks to assorted leaders of the Catholic Church, and with their Christian rivals, the Evangelicals. Yes, all these dignitaries are calling to express their solidarity, to give the thirty-three trapped men a sense of how

Chile and the world are behind them. Urzúa is a generous and grateful
man, and never complains about being treated like a caged celebrity
forced to speak to anyone his surface handlers see fit to bring to the
phone, even though he'd be well within his rights to do so. Urzúa isn't a
man to pick a fight, either, not with so much responsibility already on
his shoulders. He's the miners' liaison to the psychologists, the engineers,
the doctors, and, most important, to Sougarret and the minister of mining and the president. Finally, he realizes: "I had to start delegating
more." He puts Samuel Avalos, aka CD, in charge of the thermome-
ters and the hoses that have begun to pump a bit of fresh air into the
mine. (Avalos notes that when the air isn't working, and during certain
phases of drilling, the temperature rises to as high as 50° Celsius [122°
Fahrenheit] and the humidity reaches 95 percent). When a fiber-optic
line is lowered down the shaft, Urzúa gets two of the younger and more
tech-savvy men to make the necessary connections and to be the new
communications team: Ariel Ticona, twenty-nine, whose wife is about
to give birth to their baby, and Pedro Cortez, twenty-six. Then the new
communications team quickly gets into a fight with Mario Sepúlveda,
because Mario feels he should be able to pick up the phone and talk to
the surface whenever he wants. "Perri almost got into a scuffle with
Ariel," Víctor Segovia notes in his diary. Ticona and Cortez both quit
their new jobs and retreat to the Refuge, until Urzúa goes and finds
them and convinces them to return.

The new fiber-optic line is quickly put to use. On September 7,
Cortez and Ticona and others connect it to the portable Samsung projector, which casts a television image onto a white sheet: The national soccer
team of Chile is playing the team from Ukraine in an international
friendly. Live, from Kiev. The Chilean team poses on the field for a
picture before kickoff, wearing shirts that say FUERZA MINEROS. Almost
all of the thirty-three men gather before the makeshift screen to watch
the game, and most put on the matching red shirts they've been sent
from the top. They are filmed while they watch, in a black-and-white
video the Chilean government will distribute later to the global media.
Franklin Lobos, who once wore the Chilean national team jersey as a
player, offers a commentary on the game to Chilean television, which
only adds to the amusing weirdness of what will become a light feature,
related by news anchors with a gentle smile on their lips, about a group

of working stiffs trapped under two thousand feet of stone doing the most normal-guy thing a guy can do: watch soccer. The men smile and wave at the camera, and in their identical shirts they resemble a crew sent on a mission to the center of the Earth. The men are unwitting subjects of global entertainment, but they don't complain, though Víctor Segovia decides not to join his colleagues watching the game because he doesn't want people on the surface to think that everything is okay inside their cavern-prison when it really isn't. Eventually, more of the men will rebel against the idea that they are fish in a fishbowl: For a few hours they cover the camera that broadcasts a continuous image to the surface at the site where the *palomas* are unloaded.

Up on the surface, at least some of the family members are also resisting the sense that the collapse of the San José Mine is becoming a media- and celebrity-driven event. In addition to the politicians, the diplomats, and the philanthropists, actors and musicians are showing up at Camp Esperanza, too, like the Stetson-wearing Chilean cumbia-ranchera group Los Charros de Lumaco. The families are invited to pile into buses and attend a comedy show in Copiapó, and later there will be a lingerie giveaway for the wives and girlfriends. The musicians, comedians, actors, and purveyors of women's underwear have come with no other intent than to prop up the families' spirits, but Carmen Berríos, the wife of Luis Urzúa, has no use for any of them.

"People come from the outside, 'artists,' they say, who don't interest me personally," she writes to her husband in the mine. "You know me, and know what I think about that." Carmen ignores both the celebrities and the reporters who want to transform *her* into a celebrity. Luis Urzúa is, along with Mario Sepúlveda and Yonni Barrios, one of the miners getting the most ink and airtime, and the reporters want to make Carmen and her two children stand-ins for the leader down below. "The family made a promise: Noelia, Luis, and I will not respond to the 'idiotic' questions of the press," she writes. "And your family, your mother, brothers, cousins, etc.: They also know how we feel about these reporters and they have to respect what we've decided. Only when the miners are rescued will we talk to them. For that reason, you will never find us in any newspaper."

Down below, Luis is devouring his wife's letters. Carmen is herself, she hasn't changed, and her daily letters are a measure of sanity amid so much around him that is absurd and frightening. He wants her to write more, because when she does things feel normal, for an instant, as if he were home at the dinner table listening to her. Carmen can't help but note the irony: "You always said I talked too much," she writes. Carmen tells him again and again to rely on their shared faith ("Are you reading that prayer book I sent you?") and to keep focused on what's important, the rescue and the safety of the thirty-two men who are his responsibility, and not the fame and wealth that might be coming his way. Not once does she mention the Farkas millions, and when he finally asks her about it, she writes back to say: Don't worry about that, you need to focus on the rescue, because it's ridiculous to talk about money when your life is still in danger. Luis trusts Carmen more than anyone, and she's also become his eyes and ears on the surface. It's one thing to have these officials telling him the Chilean government is doing everything it can to rescue them, but it's quite another to read it in Carmen's own words. "You can't imagine the great deployment of machines, of workers . . . of huge floodlights, of shipping containers and of machines carving out roads over the hill to the south, north, west, and east," she writes. "There are trucks moving and taking out earth, and huge trucks that go up every day filled with water, 5,000 liters at a time. We hear the buzz of the generator engines that keep the floodlights on: In every corner of the mountain there are huge floodlights, of the kind you've probably seen before only in Punta del Cobre [one of the biggest mines in Chile] or some other place, but never in the San José."

Carmen writes to her husband, above all, to remind him that he is loved. She writes a poem and sends it to him. "Generous, simple, suffering miner / drinking from the wounded innards of the Earth / looking without seeing with your tired eyes. / I say miner and I say your name / miner of coal, of dust, of minerals / your hands wizened by early winters / that know the shovel, the carved rock, and the pick." Sometimes, her letters are playful and girlish, as if they'd just met a few weeks earlier, and hadn't been married for twenty years. "Have you thought about me, missed me, or have you forgotten the kind of perfume I use?" she asks in one. (Luis has not forgotten: For several days after the collapse, he smelled her perfume on the seat of the pickup truck when he slept—it had hitchhiked from home on his clothing and affixed itself there somehow.) More often, her letters convey a devotion that is mature, romantic, and enduring. "Don't forget me," she writes. "Remember the good things that I gave you, and the bad things, too, which were just a few. Because we will see each other again, to begin anew, just like the first time." She concludes that letter with a promise: "I will wait for you forever." *Te espero por siempre.* When Luis holds Carmen's letters in the white pickup truck that is his underground office he finds it a little easier to believe that he will escape this collapsing mountain with his sanity intact and that he could make it to December, or even January, underground with all these angry men.

Most of the families of the thirty-three men know it's their duty to convey calm and a sense of domestic order to the men below, though this gets much harder to do after the new fiber-optic line from the surface is connected to a videoconferencing system. Omar Reygadas, the white-haired scoop operator who used a flame to follow a breeze to the bottom of the mine, can now see the face of his handsome adult son, Omar Jr., on that screen. For both men it's an act of will to keep away the flood of tears that will follow as soon as their allotted time to speak is over and the connection goes dark. They feel shattered by this moment they are living, by their smallness before the mountain of stone that's trapped Omar, which somehow feels more real and daunting now that they can see and speak to each other. "I wanted to cry, but I didn't, and of course I found out later my family wanted to cry, too," Omar Sr. says. Father and son don't want their tears to add to each other's burden. Instead, Omar Jr. says: "We always had the faith that you were alive, that God was going to protect you down there." Omar asks about the elderly aunt who raised him, because she has diabetes and Omar has always cared for her. She's okay, his son says. What about the bills, the rent? "*Viejo*, don't you worry about anything, because I've got all your bills paid, on time." Omar Sr. rents part of his house and the tenants are all paying on time, too, Omar Jr. says. "Everyone is behaving well, helping out, they've all paid. Don't worry about anything besides taking care of yourself inside," he says. His other relatives are equally upbeat.

"They transmit good feelings to me," Omar Sr. remembers. "Their good vibes." *Su buena vibra.* In his letters he jokes with his family about wanting his favorite meal, steak and avocado, which is just another way of saying: I'm still the same old man who left for work on August 5, the man who works hard and eats his *corazón con palta* afterward. "In the letters we wrote to each other, there was never a misunderstanding, never a fight with my children or a complaint. To the contrary, we always had a very healthy conversation. With my wife it was the same. She'd go up [to the mine], turn in her letters, and go back to town to attend to her work. It was just pure love letters, that's all." In the letters from the surface, Omar Reygadas can feel the rhythms of normal life that are waiting for him outside. His job in the caverns of the San José is to do his share of the little work there is to be done, and also to recuperate and rest, despite the darkness, the rumbling of the mountain, and the complaints of so many of his coworkers. Mostly, he reads on his cot, by the entrance to the Refuge, with a box as his nightstand. His relatives have sent him several volumes in a series of pulp cowboy novels by the Spanish writer Marcial Lafuente Estefania, who writes books about Texas Rangers and bandits with titles such as *El caballero de Alabama* (The Alabama Gentleman) and *El Capitán "Plomo"* (Captain "Lead"). The hero on the cover is wearing the same wide-brimmed, flattop gambler Clint Eastwood wore in his spaghetti Westerns. But the book Omar finds most enjoyable is Paulo Coelho's *The Alchemist*. Several million people have read this book, and now Omar does, too, on his inflatable cot down at Level 90. He reads the story about a shepherd boy crossing the desert, and its many uplifting aphorisms: ". . . when you want something, all the universe conspires in helping you to achieve it."

As Omar reads, forty-two men are crossing the Atacama Desert, headed for the San José, conspiring to achieve his liberation from the mine. They are not shepherds, but truck drivers, hauling more equipment to try to pull thirty-three men out of the mountain. The huge oil-drilling rig that will be used for the Plan C rescue includes a 45-meter (147-feet) tower that has been disassembled into component parts. As the sun rises on September 9, the slow-moving caravan of trucks transporting it is just twenty-four hours away.

Jessica Chilla, who received a long, unexpected embrace from her common-law husband, Darío Segovia, on August 5, is another one of the family members who's avoiding the press, mostly because she's afraid she won't be able to control her emotions if someone puts her in front of a camera. "If anyone saw me, I wanted them to see the same Jessica who was going to greet her husband when he came out of the mine," she says. "The Jessica with her shoulders held high, waiting for him, giving him strength. Because anyone who says they heard me cry— well, they didn't hear me cry. I had to be strong for him, so that he could get back on his feet when he came out." Jessica wants Darío to see that she's "*super bien*." That's why she makes the mistake of sending him a picture in which she looks cheerful as she stands with his sister María and the members of Los Charros de Lumaco.

The picture was taken during Los Charros' visit to Camp Esperanza, and Darío receives it in a letter delivered through the *paloma*. Is this some kind of joke, he wonders? The photograph shows his wife, up there in the midday sunshine, standing alongside six clean-shaven and hand- some men dressed in identical black Stetson hats and black shirts with embroidered white flowers. These men all look younger and better fed than he does, and some of them actually have their arms around Jessica.

"Why did you send me this?" Darío writes as he sends the photo- graph back to her. "I don't want to see any *huevones*. And much less do I want to see musicians touching you."

"He was always jealous," Jessica says. "And now, being stuck down there, he was doubly jealous."

Jessica is doubly hurt—by Darío's angry words about the photo- graph, and by the implication that she's living it up on the surface while he's suffering down below. No, they aren't living a party in the camp. Sleeping in a tent isn't fun. It's cold at night and she's exhausted from the effort of keeping up her spirits for Darío's sake, amid people who don't seem to understand that his life is still in danger. There's no guar- antee Darío or any of the other thirty-two men are going to get out alive, and yet all around her in Camp Esperanza there are people who treat that growing gathering of relatives, rescuers, and reporters as if it were some sort of neighborhood street festival. "There were people showing up just to get a free meal, since there was stuff there that had been do- nated by the Jumbo supermarkets. You could eat whatever you wanted. They'd give you chocolates, boxes of tea, and there was a little cart where you'd line up and they'd give you French fries, hot dogs, tortillas."

Many of the people of Copiapó and the surrounding towns in this mining region have lived a very austere and hard-fought existence, giving thanks to their Creator for every piece of fresh-baked bread, second-rate cut of beef, or first-rate chicken breast that comes into their home. In Camp Esperanza they are presented with the abundance of a Jumbo *hipermercado*, all the firewood they could possibly use, and some truly excellent samples of the local seafood. Some can't help but get carried away.

Something similar is happening down below. The young miner and hard drinker Pedro Cortez is thinking what he'll do with all the money coming his way. He went in on August 5 to work one shift, and will end up getting the equivalent of an entire year's pay, or more. Before, he spent too much of his pay in the *chopperia* beer establishments in downtown Copiapó. But now he has enough money to buy a house for his parents to live in, and he tells his fellow miners that he's going to pay for his daughter, the one he all but ignored before the accident, to go to a really good private school. A lot of the younger guys around him are getting car magazines and brochures sent down to them, including Jimmy Sánchez and Carlos Barrios, and they spend many hours in their cavern prison leafing through glossy pictures of European sports cars and American pickup trucks—suddenly the multimillion-peso price tags attached to these vehicles don't seem so daunting. Pedro tells his friends he's going to buy a yellow Chevrolet Camaro, like the one he saw in the movie *Transformers*. His friend Carlos Bugueño has a more modest dream: A little Peugeot 206 would suit him just fine. Some of the older men talk about buying big delivery trucks, the kind of vehicle you can use to start your own little business.

Complicating these dreams is the fact that a lot of these prospective car buyers—including Pedro Cortez—don't have a driver's license. In Chile it's harder to get a license than in most Latin American countries, since you have to take a "theoretical" written test and you can't bribe anyone to "pass" it. Some of the thirty-three trapped men write to their relatives and ask them to send down study materials for the driver's test. Soon, Level 90 is like a little traffic school, as men ponder the Chilean vehicle code and the hidden subtleties in questions such as: Why should you reduce your speed when you drive in fog? If there's a horse rider on the highway, at what speed should you pass him? If you strike a pedestrian at 65 kilometers per hour, how likely is it he will die?

Studying for the driver's test while your life still hangs in the balance is madness. It's cash fever, and Carlos Bugueño can see that he and his fellow workers are caught up in it. "The money was starting to cloud things for us," he says. The reminders of the easy life that awaits them come from everywhere. For a few mornings, the fiber-optic link from the top brings a four-hour television show broadcast live from Santiago, *Buenos Días a Todos*. One day the "Good Morning Everyone" team announces that the government of the Dominican Republic has offered to bring all thirty-three miners and their families to a relaxing resort in that Caribbean island nation. "We're going to the beach!" someone shouts. The men haven't seen daylight in a month, most of them have never set foot outside Chile, a few have never traveled beyond the fringes of the Atacama Desert, but one day soon they will all visit this new and heavenly place of hot sand and turquoise water together.

"It was surreal," Luis Urzúa says. "But after a while, surreal things like that started to seem normal."

Urzúa decides that the men are spending too much time watching *Buenos Días a Todos*. They sit there for hours and neglect important work. For example, now that the men are eating regular meals, there's a lot of shit, literally, to clean down in the toilet area. And not the isolated little llama pellets of before, but rather man-size, miner-size, smelly turds, in large quantities. To get the men to clean their excrement, Urzúa calls to the surface and asks the rescuers to turn off the television in the morning. With no more *Buenos Días* from Santiago, the men finally get around to latrine duty. From then forward, the television is on only during the afternoon, for soccer games involving Chile's most popular teams, La Universidad de Chile and Colo-Colo, and for movies "to keep us calm and to keep us from complaining," one of the miners says.

Not everyone trapped below is taking it easy while waiting to be rescued. In this first week of September, Víctor Segovia notes in his diary an odd sight: Edison Peña is running through the mine. He's taken a pair of boots and cut them down to ankle height, and uses those to run up and down the dark passageways, alone with the beam of light on his helmet and the sound of his breathing through the thick air. Edison has long been the eccentric of the A shift: He used to walk alone in the mine, and sang Elvis songs in the Refuge, and when they were starving he performed those morbid death skits with Mario Sepúlveda. But running for exercise down here in hell is lunacy of a higher order. Why is Edison running? After the contact, Edison says, he was overcome with joy and gratitude. He's seen a "blue light" in the mine, the light of faith. He's promised God that he'll do something to show his devotion, and what could be more devoted than to run uphill, against a 10 percent grade, in those passageways carved from the Earth? But he's also running because he senses his body needs exercise to become well. Once he started eating real food he became painfully constipated, as did many other men. Going to the bathroom is an ordeal. "I'd go and push and push. What was coming out was really thick. Then it got stuck, somehow, and no, no, no, no. It was like trying to deliver a baby. It hurt a lot." He needs to do something with his ailing body, and he has no bike to ride, so he starts to run. Many of the guys see him and start laughing. "They'd make fun of me. No one said anything supportive. Except maybe Yonni Barrios: He was worried something would happen to me." To Florencio Avalos, it looks like Edison is running "to forget things, to make himself tired so that he can sleep." Florencio also knows how dangerous it is to wander around a mine alone, and concludes that Edison, as a Chilean expression puts it, "is one plank short of a bridge" (*le falta un palo para el puente*). For Edison, running through those corridors where a falling slab might kill you is his way of saying he's going to stand up in the face of adversity. Later, he will have some running shoes sent down, from a certain globally popular brand, and then a pair of neoprene slippers. Running liberates his mind, but it also reminds him where he is and what he's been through. "I felt completely alone," he says.

While Edison Peña runs, other men drill above him. The Plan B drill has advanced 200 meters by September 9. The hardness of the diorite, the profound depth, and the angle and curve in the original, smaller borehole it's following cause the drill bits to wear out more quickly than they would otherwise. They have to change bits every twelve hours. The drilling slows from 20 meters an hour to as slow as 4. In the drilling team there are Americans from Center Rock Inc. and Driller Supply, and Chileans from the local mining company Geotec, and others. All of them, working together, are pushing their bodies and the drill past their limits. They're so eager to reach those living souls that they succumb to a phenomenon that Laurence Golborne and André Sougarret have seen before: Like the man who kept drilling long past the lowest level of the mine when the rescuers were first searching for the miners, they can't "let go" of the hole. In their anxiety to reach the men at the bottom of the shaft, they drill when they shouldn't. Unfortunately, while the men on the crews can summon the will to work past the point of exhaustion, the metal in the drill bit must still obey the laws of physics, and it finally and inevitably shatters, at a depth of 262 meters (860 feet), as revealed by a sudden drop in pressure in the T130 drill, and some schizophrenic behavior from the torque gauges. The crews raise the massive hammer to the top, and lower a camera to discover a basketball-size chunk of the drill bit is stuck in the hole, rendering the shaft useless.

Not long afterward, the Plan A drill suffers a hydraulic problem and also shuts down. The reassuring sound of drilling traveling through stone to the trapped miners below stops, and in the silence that follows, the miners begin to feel more alone, abandoned, and desperate than they have since the first drill bit broke through above the Refuge. They write letters and make telephone calls to the surface demanding to know what's going on, and they soon discover they might be stuck until December after all.

Edison Peña returns to one of the passageways alone, and allows himself to slip deeper into loneliness than any of his colleagues, listening to the suddenly louder and clearer sound of his chopped-up "running boots" striking the mine floor one stride after another. Florencio Avalos, Luis Urzúa's youthful second-in-command, decides he's tired of sitting around waiting to be rescued. He gathers rope and other tools that can be used for climbing, and with three other men heads up toward the gray curtain of stone that's blocking their way out.

15

SAINTS, STATUES, SATAN

Before leaving on his escape expedition, Florencio Avalos calls to the surface and talks to his old friend and compadre Pablo Ramirez. I'm going to try to find a way out through the chimneys, Florencio says. Ramirez tries to talk him out of it, of course, but Florencio won't be dissuaded. With his brother Renán, and with Carlos Barrios and Richard Villarroel, Florencio drives the kilometer up to Level 190 and the chimney closest to the site of the collapse. Their plan is to retrace the path Mario Sepúlveda and Raúl Bustos followed on that first night underground, up the chimney to the next level, and perhaps to another chimney after that. They start up the jumbo and its attached cherry-picker basket, and they begin to climb.

On the surface, the rescue team isn't ready to give up the Plan B hole. If they can get that chunk of metal out, they can resume drilling. They lower a magnet into the hole, but they fail to lift the shattered drill bit. On that same day, the American driller Jeff Hart arrives at the mine site following his long journey from Afghanistan. He's there for the next, final phase of the Plan B drilling, a mission that's on hold until the currently blocked hole can be unblocked, or a new Plan B drill can be started.

In the midst of the silence and the waiting, Carmen Berríos receives a letter from her husband, Luis Urzúa. He says the men are desperate because they can't hear any drills. "The rescuers have worked hard for you," she writes back. "Because God is with them. But if you all down there stop believing and stop praying it will all be for naught. Don't you think? . . . Now, if you don't hear the machines drilling, it's not because they've left. Just have faith and don't surrender to desperation. I write this because I want you to understand that a single objective motivates all the people involved in the rescue: getting you out of there."

Before nine on the foggy morning of September 10, the trucks carrying the parts for the Plan C drill arrive after their long drive across the Atacama. They drive slowly up the narrow road leading to the San José property. It's cold and the mood among the family members gathered outside the mine is subdued, though several wave Chilean flags and a few manage another chant of "Chi-chi-chi, le-le-le." While the trucks ahead of him park, one driver stops his vehicle, and as he waits near the gate, a television news crew stops to talk to him: "We've arrived, with many sacrifices, after crossing the desert," the driver says, and he's visibly moved to be at the mine, where his countrymen and so many people around the world have focused their hopes. "But here we are, with our hearts big, like all Chileans."

Above Level 190, Florencio Avalos and his three companions are summoning the courage to crawl up the chimney. They reach the opening to the next level of the Ramp, and walk toward the second, higher curtain of gray stone blocking the road to the surface. Florencio and the other miners begin to clear out small boulders at the site of this collapse, rocks that are on top of a huge reclining stone. Very soon, he's cleared a space big enough to squeeze through, crawling like a cat. "I'm going in there," he says, and Carlos and Renán and Richard all tell him it's too dangerous. But Florencio squeezes through, and as he does so he sees a vast, open black space that swallows up the beam from his lamp. He crawls toward this precipice and loosens a rock, which falls into the blackness and lands with a crackling clap about two or three seconds later; his experience as a miner tells him the rock has fallen some 30 or 40 meters, roughly the height of a building that's ten or twelve stories tall. He realizes he's near some sort of new, interior *rajo*, or cavern. To advance farther, he ties a rope around his waist and passes it back to his colleagues, "because I knew that any wrong move and I might fall." He's able to crawl out of the crack and stand on a rock overlooking this cavern. "I shone my lamp and saw nothing but rocks, in this enormous space, and I thought, *We can get out through here.* I knew that from that point it was just another thirty meters up to a place where it was clear, and what I could see upward was thirty meters." But Florencio also can see that the crack he's just squeezed through is too narrow, and the remaining climb is too strenuous, for all of the men to make it. The bigger and older men would still be stranded. "At most, fifteen or twenty of us were going to be able to make it out this way. Luis Urzúa wouldn't be able to make it. Franklin Lobos wouldn't either, or José Henríquez or Jorge Galleguillos."

When the escape party returns back down to the Refuge, Florencio learns that André Sougarret has been trying to reach him. Don't try that again, he says. It's simply too unstable and too dangerous. Florencio has set eyes upon the new chasm created by the collapse and explosion of the skyscraper-size chunk of diorite that destroyed the mine on August 5. The crumbling mountain is still spitting rockfalls every few days or hours, and Florencio is fortunate to have seen this chasm, and to have stood inside it, without being seriously injured.

At 10:00 p.m. on September 13, with a group of engineers, mechanics, and drillers still trying to rescue the Plan B hole, the Virgin Mary arrives at the San José Mine. She's made of wood, a newly sculpted representation of the Virgen del Carmen, the patron of Chile and spiritual guide to the soldiers who fought in the War of Independence against Spain. The Ecuadorian artist Ricardo Villalba carved her, under commission from Pope Benedict XVI, who's blessed her and given her to Chile to mark the country's bicentennial. She's been touring the mining cities and towns of northern Chile, and since August 5 thousands have asked her to intercede on behalf of the thirty-three trapped men. As the Virgin is carried onto the mine property inside a glass case, several women gather below her with candles whose yellow flames are protected from the Atacama wind inside holders fashioned from discarded cups and bottles. A flickering yellow light glows through the plastic skins of these humble receptacles, painting the faces of the faithful with a glow that's warmer and kinder than the stark gray of the flood lamps that hover over the camp. Following the liturgy led by the bishop of Copiapó, Monsignor Gaspar Quintana, the women whisper prayers, and as they do, some allow the hot wax of white candles to drip over their fingers, until their wind-chapped hands themselves begin to re-semble weeping wax sculptures. They pray for the Virgin to remove the obstacles that are keeping those thirty-three workingmen in darkness, under the ground upon which they are standing. From behind the glass, the Virgin sees their devotion and looks down upon them with the faint, fixed beatific smile that the sculptor Villalba has given her.

News of the presence of the Virgin Mary on the surface soon arrives below. For the Catholics, the power of the mother of God can be sum-moned to Earth, and sometimes it takes concrete form in an object said to have been created by the hand of God: for example, as in the Virgin of Candelaria in Copiapó, a tiny stone sculpture said to have appeared, miraculously, to an eighteenth-century mule driver seeking shelter from a storm in the nearby mountains. People venerate these objects because they feel closer to God in their presence. Now several of the Catholics trapped in the San José Mine will credit the Virgen del Carmen for the fortuitous event that follows mere hours after she's left the mine: the rescue of the Plan B hole. The drillers and engineers at the Plan B site have lowered a metal "spider" into the hole and managed to retrieve a 26-pound chunk of metal stuck inside, 862 feet below. The Virgin, it seems, has interceded on their behalf, and the men who are especially devout Catholics hold pictures of the mother of God, the patron of Chile, and give thanks. After five days and nights of crisis, and of prayer, the best hope for a timely rescue of the thirty-three men is back on.

After listening to all the Catholics around him boast of the powers of this or that statue or image of the Virgin, José Henríquez begins to make some casual remarks during the daily prayer about the danger of venerating images instead of venerating God. One of the miners has even done a dance to the Virgin. Henríquez finds the cult of statues at once quaint and offensive. It's one of the Ten Commandments, after all: Thou shalt not make unto thee any graven images. On Mount Sinai, God came to Moses and said not to bow before such objects. Eventually, Henríquez will expound on his beliefs in a way several miners find insulting. "To a certain point, Don José wanted to impose his religion on us," Omar Reygadas says. "He started to renounce the saints. I don't believe in the saints either, but I respect all religions. The people going to the prayers were from different religions, some were even nonbelievers who just wanted to pray. And there were a lot of people devoted to the Virgin of Candelaria, who is the one they say takes care of the miners. So when Don José started to speak out against the saints and the adoration of images, those people were offended." Henríquez says, "I didn't attack anyone. Did I make a commentary? Yes. Because it's there in the Word: Don't worship images."

Víctor Segovia, who was never religious before, loves to attend the informal underground church at which José Henríquez is pastor. But he, too, is put off by the direction the services are taking. One day in September he describes going to the daily service and seeing Osman Araya, now fully recuperated from starvation, slip into the holy trance of an inspired Evangelical pastor, raising his arms up in the air because he's really feeling the Lord. "I am no longer enjoying the noon prayer as much because Osman has started to scream and cry when he prays, and that reminds me of those churches where they cry and jump and scream," Víctor writes. To Víctor it looks theatrical and strange, though he will continue going to the daily prayers led by José and Osman after others have dropped out.

Omar Reygadas also attends the prayers and notes those who are missing: "Franklin Lobos started praying by himself. Others would step to the side and do their own prayers. And some just forgot about praying and would listen to music."

For Mario Sepúlveda, who first issued the call to prayer five weeks earlier, the absence of his fellow miners at those holy sessions is another blow. Once all thirty-three men prayed together, but eventually, fewer than half a dozen men will stand with the Pastor to hear the word of

God. Mario can see that the brotherhood that kept them together is fall-ing apart, and the distress this causes him leads him to go walking downhill, into the deeper recesses of the mine, down to Level 44. It's one of the most recently excavated corners of the mine, and in a moun-tain filled with perils it's an especially dangerous place, and also hotter and more dank, thanks to the water filling up in a pool there. The un-derground pond and the large open space adds to the mystical feel of Level 44. Mario has claimed this fetid corner of the mine for himself, calling it his "sacred place" (*lugar sagrado*), and he's moved around some stones to build a shrine and speaker's podium there. He goes to Level 44 alone to read Bible verses, and to practice public speaking. On the video the miners sent to the surface, Mario looked into a camera and spoke to the entire world; in this place he reads Bible verses and speaks to multitudes that exist only in his imagination. He's practicing because he can see that his future, once he leaves this mine, will be as an orator, traveling the world to speak of God and the strength and goodness of the Chilean workingman. In his solitary speeches he tells stories about riding bikes with his son, Francisco, and about tending to the horses he owns. The sound of his voice echoes back to him in that stone chamber. But now, on September 11, his thirty-seventh day under-ground, he's going down to this empty gallery cut from the rock, this personal auditorium of his, not to speak as much as to pray and collect his thoughts, and to ask God what can be done about bringing together the increasingly divided and angry men living in caverns higher up in the mountain. Mario knows it's his thirty-seventh day underground be-cause he's been keeping a tally on his helmet since the first day. It was after he made the twenty-second mark that the men started to turn on themselves, and now on day thirty-seven, "crying, I went down there, asking God to make me stronger, asking God to do his will with us. Because the insect, the devil, was circling us."

The devil is present in the mine, taking form in all the greed, the misunderstanding, the envy, and the betrayals among the men. He be-lieves that the devil has come from the surface, attaching himself to those letters, the offers of money and fame, to pit them against one another.

Mario begins to pray: "My Lord, protect us and get this insect out of our minds. The devil has entered the soul of each and every one of us. Have pity on us, and make us as we were before. And my Lord, you can start with me, because the truth is I'm afraid of evil."

Mario is speaking these words when he hears a tremendous crash. A huge stone slab in that unstable cavern has broken off from one of the walls, some ten feet away, as big and as lethal as the one that maimed Gino Cortés. To see a slab of rock fall is not an unusual event in the mine, but having that stone crash nearby when he's talking to God about the devil causes Mario to recoil in shock and fear, and at that same instant he feels the presence of someone just behind him, a kind of hot breath that strikes him on the back of the neck. "Who goes there?" he shouts. He turns around, swinging his lamp, and shines it on the pool of water, and as he does so he sees a pair of startled, half-crazed eyes looking back at him—his own eyes, reflected by the water. He sees the face of his own fear, and it shocks him more than anything else he's seen in the mine in the thirty-seven days he's spent down below.

"¡Diablo!" he shouts into the blackness. He can feel the devil trying to grab hold of him. Suddenly, evil isn't just an idea, it's a presence lurking down there in Level 44, hovering over the water. "You'll never take me, I'll never be your son!" The crashing stone, the image of his own face in the water, and the hot breath on his neck all send Mario into a crazed state of mind in which he truly believes he's at war with an evil being. He scrambles in the mud in search of rocks and begins to throw them at the darkness, at that thing down there in the cave that's trying to get inside his skin. "I'm never going to be your son! ¡La concha de tu madre!" He throws rocks against the walls of the cave, and then he runs away, uphill, three-quarters of a mile toward Level 90 and the living souls trapped there, waiting to be rescued.

When Mario reaches the others they see him with his clothes and face covered in mud, as if he'd been wrestling with someone down below.

"What happened to you?" they ask.

"I was fighting the devil," Mario says.

Some of the men laugh, but others don't, because just about everyone who's worked at the mine long enough will have seen or felt the devil living down there at one time or another. A Chilean mining legend has it that Satan lives in all gold mines, and gold is precisely what they were digging out of the stone, down there, in those caverns at the very bottom of the mountain. The men dug out tons of rock to get at a few precious ounces of gold, and in so doing they weakened the mountain and transformed it into a mine whose walls can burst without warning. The men of the San José have seen rock explode, and it's put the fear of God and the fear of the devil into them. Sometime after Mario Sepúlveda's fight with the devil, there's another collapse down at Level 44. A chunk of rock weighing more than a ton breaks away from the ceiling of the cavern with another huge crash, and the place where Mario built his auditorium and his chapel is declared off-limits.

16

INDEPENDENCE DAY

When he entered the mine on August 5, Ariel Ticona knew that his wife was due to have their third child, a girl, on September 18, Chilean Independence Day. For the first seventeen days he was trapped underground, he told himself he needed to stay alive so that he could rise to the surface and see the girl whose name he and his wife had already agreed would be Carolina Elizabeth. Perhaps it was the desire to see his daughter that led him to privately ration the extra cookies he was given by Víctor Zamora after the first night's raid on the food stores—four cookies that he ate, secretly, over the course of that first week. After the miners had been discovered on the seventeenth day, Ariel held on to the idea that he would be rescued in time to see Carolina born, and that he would be able to fulfill a promise he'd made to his wife: For this baby, unlike the previous two, he would be inside the maternity room with her. Ariel is twenty-nine years old, and he admits to being more mature today than he was after first becoming a father. After a man has a couple of kids he has a greater appreciation of the domestic labor that a family is built upon, and with his wife's third pregnancy he had tried to be more helpful. He was there to help her do the laundry, for example, and he had hoped to be there for her final moments of labor, to hold her hand and make her stronger.

Ariel has resigned himself to missing his daughter's birth, but in the meantime, he's had an epiphany. After talking to his family via the video link, and seeing the images from the surface of the camp where thirty-three families and hundreds of rescuers have all gathered, he decides his daughter should be called Esperanza. On September 14, Esperanza comes into the world, at a hospital in Copiapó. His sister-in-law takes a camera into the delivery room and the Chilean channel Megavision prepares a video of the birth with music in the background. But Ariel doesn't see it. Esperanza has been delivered via cesarean section, and the psychologists have decided, according to media reports, that Ariel should be spared the trauma of seeing a surgical procedure while he's still trapped underground. Instead, Ariel sees a heavily edited video transmitted via the fiber-optic link to the big screen below. The other miners think Ariel should see his daughter's birth in private, and they leave him alone with the screen. He sees blue-clad doctors standing over his wife, and then the video cuts to one of the doctors holding his new daughter, and he sees her with wet, matted hair and closed eyes next to his smiling, exhausted wife. Controlled by men on the surface, the same two minutes of video plays over and over again in a loop. The quality, however, isn't sharp enough for Ariel to make out if Esperanza looks like him or his wife. No one before in human history has witnessed the birth of his daughter while trapped in a stone cavern, and when I later point this out and ask Ariel what it was like to first cast his eyes on his daughter, he says: "I don't know what I felt. If it was emotion, or happiness, or what." After speaking to Ariel's brother, the world's newspapers will report that Ariel wept copious tears upon hearing the news. They report his daughter's vital statistics, too: 3.05 kilos (6 pounds, 11 ounces), 48 centimeters (19 inches), born at 12:20 p.m. These figures are tossed into their stories alongside the latest statistics on the drills trying to reach the trapped men. The Plan B drill has advanced 368 meters. The Plan A drill 300 meters. The Plan C drill will begin work in seven days.

The drill that breaks through first will be used to lower an escape capsule to the thirty-three men. The Chilean navy begins to build that capsule—coincidentally in the same shipyard where the mechanic Raúl Bustos repaired engines until the tsunami hit. A two-minute walk in the vast ASMAR shipyard separates the small workshop where Bustos worked from the machine shops where the escape capsule will be assembled. The interior walls of most of the buildings in the shipyard in Talcahuano still have seven-foot-high watermarks from the ocean water that swept through six months earlier, and parts of the vast complex are still waterlogged. But the navy has cleaned out all the dead fish, removed the grounded vessels, and gotten the shipyard working again. Now the team of naval engineers and machinists gets to work building what their colleagues at NASA have named—following a typically North American obsession with acronyms—the EV, or Escape Vehicle. The Chileans have received a twelve-page memo from NASA detailing the space agency's recommended specifications for such a craft: "EV . . . shall have portable oxygen tanks of sufficient size . . . to provide medical grade oxygen at the rate of 6 liters per minute for up to 2–4 hours . . . EV shall be configured such that occupant is able to move at least one hand to his face." But the design the Chileans come up with is entirely their own (they will soon consider patenting it), and on September 12 the government announces its basic parameters to the media. Built from steel plates, the Escape Vehicle will have an exterior diameter of 54 centimeters (21.25 inches), will be no taller than 2.5 meters (8 feet), and will weigh approximately 250 kilos (550 pounds) when empty. It will have the oxygen supply recommended by NASA, and also a roof built to resist objects falling from great heights, and it will travel up and down with wheels that keep the Escape Vehicle's steel shell from striking the walls of the shaft as it rises to the surface. (Those retractable rubber wheels will eventually be provided by an Italian firm.) Should the man traveling inside lose consciousness, a harness will keep him standing up.

A few days later, the Chilean government releases drawings of the proposed capsule, painted the colors of the national flag, and emblazoned with a name: Fénix, or Phoenix in English. Phoenix is a minor constellation in the southern sky, a group of stars in the shape of a triangle and a diamond, two simple shapes that, when joined together, form the bird that rises from the ashes in Greek mythology. The name has an obvious rhetorical purpose for the Chilean government: Chile itself is a country that's rising from the ashes. With this capsule Chilean workingmen and Chilean technology and Chilean faith are going to pull off a daring rescue that will fill a people with hope just months after a

disastrous earthquake and tsunami claimed the lives of so many inno-
cents and sent the national mood into a funk. Lifting thirty-three men
up from the bowels of the Earth in a Phoenix the colors of the flag also
suggests how the government wants the rescue to be remembered: as a
heroic, nation-defining myth, with real Chilean workingmen cast in
the leading roles.

In Greek mythology, however, even the gods are imperfect, imbued
with vanity, courage, pride, familial love, vindictiveness, and other all-
too-human qualities that can also be found among the men living inside
the broken San José Mine.

In the days before September 18, Chilean Independence Day, the question arises: How will the thirty-three Chilean patriots trapped in the
San José Mine celebrate? Several of the leaders of the rescue team on
the surface want to send the men wine. It's the biggest holiday of the
year, after all, celebrated with family feasting and drink, and seeing
these living symbols of national pride having a little glass in their mountain prison will make all of Chile feel good. "I wanted to send them
wine, too," the psychologist, Iturra, says. "But the doctors were completely
against it." Some of the men were heavy drinkers, and they've been ab-
stinent now for more than forty days. The crisis of abstinence is over for
them: All thirty-three are now teetotalers. After thinking about it, the
psychologist agrees that wine is a bad idea. At about this time he's received a troubling reminder of the battle some of the men have had
with addiction. "One of the mothers came to me and told me, 'My son
is receiving drugs.'" The family members have been allowed to send
the miners care packages with clothes and the like, and in these packages someone has slipped in something illicit. "It was either marijuana
or cocaine, I don't know which, but it didn't really matter. I really
couldn't afford to have any men with altered states of consciousness
down there." Iturra changes the procedure by which items are packaged
and any further shipments of drugs are stopped. As to the Independence
Day wine, Iturra points out that the corridors of the mine are a work site,
and alcohol is prohibited, by law and by common sense. Down below,
the men have reached the same conclusion: We won't be needing any
wine, they say, thank you very much.

The thirty-three men will, however, enjoy empanadas and a bit of steak, a simulacrum of the feast they'd be celebrating on the surface. They prepare for Independence Day by writing a poem to the president. "Even that almost caused a fight between Perri and Edison because they had different ideas about the poem," Víctor Segovia writes in his diary on September 16. "Then Zamora jumped in and there was a very heated argument: all over a poem for the bicentennial. Hahaha."

But the bad feelings don't last long, because the preparations for the bicentennial coincide with excellent news from the surface: The second stage of the Plan B rescue is nearly complete. On the morning of September 17, the drill breaks through. A 17-inch hole now links the trapped men with the surface. Once this borehole is widened to 28 inches, the men will be free. If all goes well, that might be in just a few weeks. "This is happening really fast and it's making us very happy," Víctor Segovia writes in his diary. The next morning, Independence Day, finds most of the men getting haircuts, taking baths, and changing into clean clothes, "as if we were prisoners and it was visiting day."

Outside, photographs of the miners are displayed again and again during public bicentennial celebrations. A two-story image of the famous note *"Estamos bien en el Refugio"* is projected upon La Moneda palace in Santiago as part of a light show there. In the mine itself, the men eat their empanadas and down a cola drink. They raise a flag, sing the national anthem again, and watch Mario Sepúlveda perform a traditional cueca dance—which is videotaped and broadcast to all Chile.

The only miner who chooses not to participate in the festivities is Franklin Lobos, "in order to avoid having problems with some of the guys he doesn't get along with," as Víctor records in his diary. In a mine filled with men growing frustrated at their confinement, Franklin has an especially angry beast in his breast, one that hasn't stopped snarling since the mine collapsed. "I was always in a bad mood, even my friends will tell you," he says. But what most of his fellow miners don't know is that, underneath his irritable exterior, Franklin is a man softening and mellowing as each day passes, a man who believes he has learned to see himself as he truly is for the first time.

•

Before all thirty-three men trapped in the San José Mine became famous, only one of them had tasted fame. A soccer hero is a glorious thing for a young man to be, even in (or especially in) a provincial town like Copiapó. Franklin Lobos was enough of a hero that he had a nickname, and not just any nickname, but one with explosively martial and perhaps virile connotations: "The Magic Mortar," he was called, for his ability to fire missile-like free kicks into the enemy goal. He'd been named to the Chilean national team (fleetingly) in the early 1980s and worn the coveted red jersey. At about that time, which was also when he got married and started having children, he never lacked for auxiliary female companionship. "*Mujeres, mujeres, mujeres,*" he'll say, recalling those years. If he tried to go to downtown Copiapó and buy a drink, no one would let him: "Please, Franklin, it's on us! Let us buy a drink for El Mortero Mágico!"

In his thirties, Franklin's career started to fade; he held on until he was thirty-nine, later than most players. Retirement was a vacuum he could not possibly fill. "One day you have all these friends, and people want to buy you things—and then you don't. One day you have all these women at your side—and then they're gone." After his career turned to rubble, so did his marriage. He was making his wife suffer with his foul moods, his absences, so as an act of compassion he actually divorced her, "with papers and everything."

The Magic Mortar became a taxi and truck driver, and at age fifty-two he was down in the exceedingly dangerous San José Mine, moonlighting to help pay for his daughter to go to college: Carolina, the same daughter whose tears outside the entrance to the mine caused the minister of mining to weep copious public tears. Now Carolina is up there in Camp Esperanza and so is her mother, his ex-wife, Coralia. After all that he's done, Coralia is present in that weather-beaten camp, for their adult children and also for Franklin. Is she writing him love letters? "No, she's always been kind of cold that way. She didn't want to show her feelings. She would just tell me to take care of myself: that kind of thing." Her mere presence, her daily vigil on behalf of her cheating but now caged ex-husband, is a kind of love poem. Finally, his nephews started to lobby on her behalf, telling him: Uncle Franklin, Coralia is here every day! She really cares about you. So now Franklin Lobos, the

Magic Mortar, is considering a step that would have been unthinkable when he reported to work on August 5: getting back together with his ex-wife.

Franklin is contemplating a return to an earlier, simpler, and nonfamous version of himself: He will be part of a couple again, with the mother of his children. As he thinks about the goodness of this personal transformation, his embrace of humility, he sees the workingmen around him getting puffed-up heads about how important they are and the glory that awaits them on the surface: They're even wearing a kind of national team jersey as they gather for their underground Independence Day celebration. It seems silly to Franklin for his fellow miners to think of themselves as national heroes when all they've done is gotten themselves trapped in a place where only the desperate and the hard up for cash go to suffer and toil. They are famous now, yes, but that heady sense of fullness that fame gives you, that sense of being at the center of everything, will disappear quicker than they could possibly imagine.

Franklin tries to speak this truth to his fellow miners, but he does so halfheartedly, because he knows the only way to learn it is to live it. Instead, he watches as his fellow miners' obsession with their public image drives them to pettiness. *I'm going to buy a Camaro. Italian television wants to talk to me. My hometown wants to give me a medal!* Franklin is especially angry with Raúl Bustos, the man from Talcahuano, for teasing Mario Sepúlveda mercilessly in the wake of the story in which Mario proclaimed himself "absolute leader." Franklin believes Raúl's own vanity is responsible for the bad feeling between the men of Level 105 and the men of the Refuge, and it's to avoid seeing Raúl (among others) that he's skipped the celebration. But up on the surface, as luck would have it, Franklin's adult daughter, Carolina, has become close to Raúl's wife, Carola, in Camp Esperanza. In a letter, she tells her father about her new friendship, how the two women talk every day and make each other stronger.

So Franklin finds Raúl at Level 105 one day, and puts an ironic arm around his shoulders and tells him: "My daughter says I have to be buddies with you now. Because she's friends up in the camp with your wife. Look, it says so in this letter." Franklin shows Raúl the letter with a grin. "But you know what, Bustos? I'll never be your friend. Never. You know

why? Because you divided the group and I'll never forgive you for it."
Franklin knows he sounds like a jerk, but he has no qualms then, or
later, about saying what he feels: "It was one of the hardest things I
said down there. But I told him to his face, I didn't say it behind his
back."

The same Franklin Lobos who is willing to reconcile with his wife
isn't ready to forgive Raúl Bustos. He will hold on to his enmity in the
days to come, even as the thirty-three men begin to work toward the day
when, God willing, they'll enter a steel capsule and rise toward the surface and the light.

On September 20 the American drillers Jeff Hart, Matt Staffel, Doug
Reeves, and Jorge Herrera and their Chilean colleagues begin to drill
the third and final drilling stage of Plan B, using the recently completed
17-inch borehole as their guide. When they're done, they'll have widened the hole to 28 inches, and thus created a passageway big enough
for the Fénix rescue capsule. If all goes well, they should be done in less
than a month. "If it takes us until Christmas," the Americans say to one
another, "we need to get out of the drilling business." To speed up the
drilling, the Plan B team allows the crushed rock produced by their
T130 drill to fall down into the mine itself. In all, the T130 will send
several thousand cubic feet of crushed stone down the pilot shaft to the
old workshop, near the spot where the mechanics used to gather to feel
the breeze that came in through the cavern of the Pit. Luis Urzúa as-
signs Juan Carlos Aguilar to lead a team of workers who use a front
loader to pick up the crushed diorite and haul it away. The engineers on
the surface have offered to send him fuel to operate the machines, but
Aguilar says he doesn't need it, because he's calculated how much is left
in the pickup trucks, the personnel truck, the jumbos, and the other
vehicles (there are sixteen in all trapped with the men), and there's more
than enough to operate the front loader for several days. The grinding,
whining noise of the loader makes the broken mine feel unbroken. This
is what the men did when the mine was still producing ore laden with
gold and copper: They lifted and carried and dumped with machines
that were like extensions of their own muscles. The sound and the feel

of small quantities of stone being scooped up and then falling are familiar and comforting to the men working below. For a short while, they are truly miners again, men toiling in a cavern, each thinking about the home that awaits him when the work is done.

When the Fénix and the Plan B passageway are complete, a rescuer will enter the capsule and journey down into the mine. His job will be to supervise, from below, the loading of the thirty-three men into the capsule—and then be the last man to leave the mine. It's a mission with enormous responsibility and great honor, and the man who gets it will remember it as the capstone of his career as a professional mining rescuer. To choose that man, the Chilean government has set up an informal competition akin to the one among American test pilots in the 1960s to select the first astronauts, as described by Tom Wolfe in his book *The Right Stuff.* The government selects sixteen finalists, all employees of three agencies that have had a hand in the rescue—the national mining company Codelco, the Chilean navy, and the GOPE elite team of the Chilean national police.

Manuel González is one of these sixteen men. He's a miner and rescuer at El Teniente mine about fifty miles south of Santiago, one of the world's largest underground mines. El Teniente has a rescue crew of sixty-two men who are a kind of volunteer fire department inside the mine. They go about their regular mine jobs—in González's case as an explosives expert and shift supervisor—but in an emergency they roll into action. A few times, his work has led him to recover the bodies of crushed miners. There are other rescuers at El Teniente who have better climbing skills than González, and they were chosen weeks ago for the first, frustrated attempt to reach the men via the ventilation chimneys. Climbers are not needed for the journey in the Fénix, however, but rather fit men with patient dispositions and strong leadership skills. With fifteen years of mining rescue experience, and with a résumé that includes work as a shift supervisor and as a professional soccer player, González qualifies on all counts. In 1984, as a member of the O'Higgins team, González played in a match against Franklin Lobos and his Cobresal team and scored the only goal of his short professional career.

Now González is one of half a dozen mining rescuers from El Teniente selected to travel to Copiapó.

When González arrives at the San José Mine he meets the other men who are candidates to journey in the Fénix and there is an instant collegiality between them. They will work to prepare one another for the rescue, and they will compete to be the man who enters the Fénix first. When the completed capsule arrives at the mine, they see that it resembles a toy spacecraft you might find in a museum for kids to play in. It has oxygen tanks, a harness, lights, and a radio, and the cigar-shaped steel skin doesn't look much different from the metal that playground merry-go-rounds are made from. The Fénix is a capsule designed to travel inside the Earth, and a bureaucrat with a more creative bent might have dubbed it the Jules Verne instead. González and the other men train inside the Fénix, which is placed inside a tube 20 meters long, and then lifted and lowered by a crane, in a simulation of the rescue to come. One at a time, the rescue candidates enter the capsule and travel up and down inside the tube over and over again, and sometimes the capsule is left stationary, and minutes are allowed to pass with the rescuer "stuck" inside this steel prison, a small taste of the torments that await them should the mountain begin to crack and rumble during their journey into the mine.

When the trapped miners enter the capsule, they'll journey upward and meet their loved ones at the surface. But who will be there to greet Yonni Barrios, the man with two households? His wife, with whom he's been corresponding feverishly? Or his girlfriend, who is also the woman he lives with (most often)? In the weeks since a drill broke through to the thirty-three men, Yonni has been living his personal life underground much as he did while he was up on the surface, in the Juan Pablo II neighborhood. On the weekends, he's been dividing his eight minutes in the videoconference booth in half: four minutes with his wife, Marta, and four with his girlfriend, Susana. "I didn't care that it was just four minutes," Susana says. "Because just one little minute was gold for me." For Susana those moments she's spent talking to her lover via fiber-optic connection have a magical and mystical quality. The first time she sees Yonni he's wearing a white coat, his uniform as the miners' medical officer. His white garb and the lights illuminating him in that cavern studio leave Susana with the impression that Yonni is in "heaven," or someplace else, far away. "He was sitting down and had this light in his eye. Like he was a Martian. He had a bright light around him, you'd only see his eyes. For a moment, I thought he was dead, and the company was playing some sort of trick on me." She begins to weep, despite the psychologist's entreaties to keep up a brave face. "You're dead!" she says to the screen. "I cried and cried and Yonni said, 'I'm alive. Chana, I'm alive. Look at me! Do you understand? I'm alive!'" After these dramatic first words, they fall into something resembling a normal conversation, and Yonni is talking to her in that gentle and uncertain voice of his, and he starts to say things that sound familiar, because they involve Marta. He explains to Susana that he'd rather not be talking to his wife in these teleconferences, but he's forced to because Marta has told him she's sick from all the stress of almost losing him, and she'll die if she doesn't speak to Yonni (literally die). Susana believes that Marta is manipulating Yonni and she forgives him, like she always does.

After the teleconferences, when Susana goes back to the home she shared with Yonni, she watches and reads as she and Yonni are cast as cheap, one-dimensional villains in a media soap opera, with Marta as the victim. A global Greek chorus of strangers loathes Susana—and she doesn't care. "My happiness was so big, I didn't even feel it. He was alive and all the stories just made me laugh. It was as if the more bad things they said about him, the more alive he was. When you're fighting against death, there's nothing that can embarrass you. Because death is such a huge thing. Let them say whatever they want, let them tie me up, let them call me 'lover.' I was the 'lover.' Sure, I'm the 'lover.' 'How many women does he have?' 'About ten women! He has more women than he has shoes!'"

17

REBIRTH

Jeff Hart wears a U.S. flag on his shirtsleeve as he works on the Plan B drill, and sometimes he wears a cloth that protrudes from the back of his white Layne Christensen helmet. Almost always, he drills with one foot on the rig, and one day the minister of mining asks why. Drilling is "a feel," Hart explains. "You have to actually be standing on the rig so you can feel what's going on." Down below, metal is rubbing and pounding against rock, and the friction is transmitted up the shaft, where Hart and his foot take note of "good vibrations" and "bad vibrations" coming from the rig. "That's how you know your bits are coming apart, or if their cutting edge is actually gone," he says. The Americans are all the more vigilant because the rock here at the San José Mine is harder than they expected. As the drilling for the final stage of the Plan B shaft goes deeper, they have to stop and change bits every 10 or 20 meters or so. "It started to get very nerve-racking, because it got very sticky," Hart says. "We're in a hole that's curving around. We have drill pipe that in our minds is in the center of the hole all the way down, but it's not, it's rubbing [against the edges] all the way down. We were losing a lot of torque value in wall rub." Hart and the three other Americans are supposed to be working twelve-hour shifts of two men each, but instead they're working sixteen- to eighteen-hour shifts, then sleeping at the mine property in tents. Drilling toward living, trapped men is infinitely more stressful than drilling toward a water table or a mineral vein. Hart feels a sense of urgency that's driven by the idea that if one of those miners gets sick and it takes an extra month to get him out, he might not make it. Hart is a parent, too, one who might spend months away from home on a job, and he wants to get all those fathers at the bottom of this hole back to their kids. Soon, the stress causes him and the Americans, like the Chileans down below, to snap at one another.

As the T130 drill gets closer, a new plague sweeps through the corridors where the thirty-three men live and sleep. First there was unrelenting heat and thunder, then water and mud and fungi; now a slow-moving cloud of dust and steam begins to flow down into the mine. For seven days beginning September 27, a misty, gritty cloud floats toward Level 105 and the Refuge and stays there. "It's 7:40 a.m. and there is dirt and steam everywhere," Víctor Segovia writes in his diary. "It's like when the fog rolls in, everything is steamed up."

The men sleep and stir in the steamy, dusty cloud and Víctor wonders if breathing this in will make him sick. There are still prayer sessions, but only a handful attend. Several of the men pass around flags and pictures they've been asked to sign, mementos of what already feels like an event in Chilean history: "If something has signatures from all thirty-three of us, it's more valuable," Víctor notes. Mostly the men write letters, and already there's a sense that they're living a time that will soon pass into memory as a written story. How will they remember it, and how will it be written? Undoubtedly, they'll recall the sacrifice of the men and women who worked to pull them out of there, and their own faith and suffering. Already, Víctor "the Poet" Zamora has penned a poem exploring those themes, and it's been published on the surface. "Keep up those spirits, comrades, we have to organize ourselves first / Come together, everyone, we have to pray," Zamora begins, describing their first underground pleas to God.* Zamora also describes the regret he felt at the idea that he was leaving his family. ". . . the only thing I thought at that moment / tell my wife and sons I'm sorry / they, with yearning, are waiting for me to arrive at the door."† He ends his message with a hopeful "We are in your hands, Chile."

---

*Arriba ese ánimo compañero, tenemos que organizarnos primero / Júntense todos, tenemos que rezar.
† . . . decirles a mi esposa e hijos que lo lament / Ellos, con ansias, esperándome en esa puerta llegar.

There is, however, another way to think of their story. Isn't it really just a series of betrayals, of men failing to see how this mine was already killing and maiming people and destined to come crashing down and kill everyone inside? The mine owners need to be held accountable, but so do the supervisors who work for them, as at least one of the men argues. After one of their daily meetings, a miner turns on the shift supervisors, Luis Urzúa and Florencio Avalos. It's because of you guys we got trapped down here, he says. If you had shut down the mine, we wouldn't have been trapped. The miner threatens to file a criminal complaint against Urzúa and Avalos when he reaches the surface for "quasi homicide," and go on the airwaves and tell the story, as he sees it, of the culpability of the supervisor and his foreman in the events of August 5.

However their story is told, there is the question of who will benefit from the telling. Let's be smart about this, several miners say, and let's not allow others to make money from our suffering, like they always do. Several of the miners have insisted, in the face of the media onslaught from above, that they need to stick to the pact of silence first suggested by Juan Illanes. Everything that happened between August 5 and August 22 belongs to the group, and no individual, Illanes has said. If they stick together, they will all share equally in whatever money there is to be made selling their story on the surface. But the temptation of individual riches is hovering over them, and sometimes it reaches down into the mine to touch them, via offers made to their relatives and relayed via the mail in the *paloma* tubes. "I've got a contract that's brilliant," Edison Peña tells Illanes, because he wants to know if he can accept it without breaking the rules. A certain athletic shoe company based in the United States has offered Peña, already famous as the "miner athlete," money for wearing its shoes when he gets out. Mario Sepúlveda approaches Illanes, too, because the pact was Illanes's idea after all, and Illanes has become a kind of legal counselor down below. Sepúlveda doesn't say he has a contract for an interview or some other media deal, but the mere fact that he's asking what, exactly, he can and can't talk about is suspicious to Illanes.

"Look, compadre," Illanes begins. "Be very careful. Because between what's yours and what's property of the group there is a very fine line . . .

If you fuck up, I'm going to put you in jail.* Let's be clear. You don't have anything down here that's just yours. Nothing. Are you going to tell me that if we threw you down here for weeks and left you all alone, completely alone, and then we came to rescue you, we'd find you as fine and dandy as you are now? Did you pull that off all by yourself? No, compadre. You made it this far because behind you there were thirty-two others."

As September winds to a close Juan Illanes and Luis Urzúa and others wonder if they should formalize their oral agreement and make it a legally binding one. "The ambitions of the families were going to lead people to break the pact," Urzúa says. "Ambition changes people." Illanes and Urzúa can see that people are leaking bits and pieces of information about their story to the press. In letters to the surface, and in discussions with the psychologist, Iturra, the men ask that a notary be sent to the mine, a person who can give the oral agreement among the thirty-three men a written form, and validate that they've all signed and agreed to it—even while they are still trapped and waiting to be rescued.

Iturra agrees to the miners' request—and gets himself in trouble with the Piñera administration for doing so. "They wanted to fire me for it," he says. The miners haven't told Iturra why they want to talk to a notary, in part because they don't want the media to figure out what they're up to, so the officials can only speculate: Have those trapped men become thirty-three ingrates planning on suing the very government that's trying to rescue them? Do they already have a movie deal with someone? Iturra tells the officials he's not concerned, because the miners are still underground and "they really can't do anything."

The notary arrives at the mine on October 2. The men discuss their plan, and the notary tells them that he can consult with an attorney and draw something up, but nothing can be legally formalized until the men reach the surface, since a notary has to be present at the signing of any document—and not watching it, from 2,100 feet away, via video-conference.

*Si vos la cagáis, yo te meto preso.

As the notary leaves the mine, the T130 drill has reached 428 meters, and is less than 100 meters from breaking through and freeing the men once and for all.

Before the capsule can reach them, however, the trapped men will have to pull off one more mining job. They're going to have to set off a blast at the very bottom of the Plan B shaft. Even when the final shaft is complete, the rescue capsule won't fit inside the mine unless the men below can remove part of the stone wall next to the shaft opening. It's a relatively routine piece of mine work, but it requires using a jackhammer to pound a hole into the stone for the explosive charges. The jackhammers need compressed air that used to be supplied in hoses from the surface until they were cut by the collapse of the mine on August 5. Jorge Galleguillos, whose job it was to maintain that air supply, now works on rigging together a series of two-inch hoses that the rescuers have sent down to supply fresh air from the surface. Having recovered from his swollen legs, he runs the hoses up to Level 135, where Víctor Segovia and Pablo Rojas eventually get enough pressure to run a jackhammer to drill eight holes in the rock.

Jorge is in the middle of this work, walking alone in the corridors near Level 105, when he crosses paths with Yonni Barrios. Yonni is feeling harried with his medical duties—he's carrying several plastic bottles with medicines shipped from the surface—and for some reason he assumes Jorge is just taking a relaxing stroll through the mine.

"Hey, asshole," Yonni says. "You're just goofing off, aren't you?"*

Jorge is tired and frustrated himself, and he responds to this insult by taking the palms of his hands, which are covered with mud from lifting up hoses, and wiping them on Yonni's white medical smock. "There, that's how lazy I've been," he mumbles. Then he slaps Yonni across the face.

Yonni drops his medicine bottles on the ground, recovers himself, and then kicks the older Jorge in the leg. After countless heated words and threats among the men of the A shift during the more than eight weeks

*Oye, culijuntos. Andáis haciendo puras huevadas no más.

they've been trapped underground, this is the first physical altercation. There is one witness, Luis Urzúa, but before he can say anything, the fight is over.

On October 9, Jeff Hart and the T130 drill crew are less than a foot away from breaking through to the workshop at Level 135 when the drill rig emits a loud pop. "It scared us all to death," Hart later says. For a few seconds the drill team considers the implications of losing the hole and all the work that went into it, but then, surprisingly, the drill just keeps going, without any noticeable change in pressure or torque. Hart will never find out exactly what caused that pop.

Minister Golborne has told the families that when the drill breaks through the crews will set off a horn. At 8:02 a.m., the wail of a siren travels across the mountain: The Plan B drill has reached its destination. In Camp Esperanza, family members call out, "To the flags!" and they rush toward a collection of flags, thirty-two Chilean and one Bolivian, located at the bottom of a scree of stones, a place where the families have met before to celebrate good news, or to face bad news together. "Our joy will be even greater when they're all taken out alive," María Segovia tells a reporter. It will take two or three days to remove the drill bit and to test the hole itself for stability and safety. Later, in a press conference, Golborne makes the observation that it's taken thirty-three days of drilling to reach the thirty-three men. The last man is scheduled to enter the capsule and leave on October 13: If you add up the digits for the date, month, and year, Golborne points out, you will also arrive at the number thirty-three.

At Level 135 the men gather to look at the hole that will take them to the surface, and celebrate with embraces, and by taking pictures with cameras their relatives have sent from the surface. They are one step closer to freedom. But Víctor Segovia is worried: The opening seems too small, and he can already sense the claustrophobic torment of being lifted up through it. A short time later, the miners pack nitrate explosives into the holes they've drilled nearby. They set off the very last explosive charges in the 121-year history of the San José Mine, a relatively small blast that will help set them free.

The next morning, several of the men awake to the sound of distant thunder transmitted through stone. A series of rock explosions can be heard, coming, perhaps, from inside the cavern that Florencio Avalos glimpsed, a storm inside the mountain like the one that trapped them on August 5. Samuel Avalos, alias CD, manages to sleep through this thunder, until Carlos Barrios kicks him in the leg and wakes him up. "Hey, CD, stand up, *huevón*! The mine is making noises. Put your helmet on. What do we do?" Pedro Cortez, the young miner who wants to buy a yellow Camaro, is woken up, too, by another of the mining veterans, Pablo Rojas.

"The mountain is cracking a lot," Pablo says.

"Yes, it's been doing that," Pedro answers. A series of strong thunderclaps, *pencazos fuertes*, have interrupted his sleep. Now they come just a few seconds apart, and they will continue for four hours, but there's nothing to be done about it, so Pedro throws himself back on his inflatable bed and tries to sleep. This infuriates Pablo, who doesn't understand how anyone can sleep when the mine sounds like it's about to collapse again. "It was funny because in the news they'd been talking about the old foxes of the mine as if they were these experts who helped all us young guys survive," Pedro later says. "And now the old guys were coming to us young guys in a panic."

Soon all the miners who sleep by the Refuge are getting up and gathering by the shaft where the men receive *paloma* shipments from the surface. The older miners look distressed and Yonni Barrios is crying. We have to talk to Sougarret and Golborne, they say. We have to tell them we can't afford to wait another two or three days. They need to come for us now. "There's this myth that the devil lives in gold mines," one of the younger miners says. Some of the men believe the rumbling is the devil, and that the devil is angry because the men are about to leave. It's a Sunday, the day the men get to talk to the surface, and in the conversations later that afternoon several miners beg their family members to tell Sougarret and Golborne to speed up this rescue, for the love of God, because the devil inside the mountain is angry and he doesn't want to let them go. But the organizers of the rescue stick to their plan: On that Sunday and the following Monday, they continue to test the stability of the borehole, and of the Fénix capsule by filling it with sand-bags and lowering it into the shaft.

The thunderclaps inside the mountain end with a huge rumble coming from below, what feels like a massive rockslide. Later, the men travel deeper into the mine and find several corridors have collapsed and caved in, including the gallery at Level 44 where Mario Sepúlveda used to pray.

On Monday, October 11, the rescuers on the surface test the Fénix capsule to a depth of 600 meters. Inside the mine, the men begin to clean up the Refuge, like travelers who want to leave an orderly home behind before undertaking a long journey. Luis Urzúa summons them to a meeting, their final one as thirty-three trapped men. He tells them they should remember how they helped one another when the mountain was collapsing, when they were starving, and how they worked together to stay alive for sixty-nine days. Several of the men then step forward to acknowledge debts and friendships. Víctor Zamora speaks of his gratitude for Mario Sepúlveda, who inspired the men in the Refuge when they were at their lowest moment. Jorge Galleguillos, the old miner and northerner who spent many days suffering from swollen legs, steps forward to thank Raúl Bustos, the mechanic and nonmining southerner. "He was always a true gentleman and I appreciate that. Yeah, he can be a bit of a grump, but he always helped me." When Galleguillos is finished, Luis Urzúa looks at him, and at Yonni Barrios, and says: "There's two men here who spent a lot of time together, who helped each other out, and who've always been friends, but who recently had a bad fight. And I think they should both step forward and shake each other's hand." With thirty-one men watching, Galleguillos and Barrios clasp hands and embrace.

Eventually, Juan Illanes steps forward and speaks in his authoritative baritone. "Since this is the last time we're all going to be together, I think we should reach an agreement that will serve us in the future," he says. From here on out, he says, a decision reached by a majority of the men on the surface should be respected by the entire group. They're all about to return to their homes, many of them in places distant from this mine, and "it will be impossible to get all of us together again. And whether you like it or not, you should all agree to respect the decisions of the group." Illanes reminds them of what they've agreed to before: They will not reveal, individually, what they suffered as a group. That story is their most precious possession, and it belongs to all of them. "I want people here to show that they're going to respect what we've agreed to, as men. It's not just about the group, it's about how much you love yourself. Because if you love yourself, you will the defend the rights of others." Illanes loses his composure as he reaches the end of this speech, because he's asking the men to believe in a promise expressed in words, a weighty and sacred idea that is also as fleeting and passing as the breath with which those words are spoken. He's asking them to remain loyal to abstract notions of honor and solidarity, against the temptations that await them—real fame and real money, up there in the crazy, unfettered world of the surface. One of the miners steps forward to express his dissent—Esteban Rojas says he doesn't trust Illanes and doesn't agree with much of what he's said. There are murmurs of agreement with Rojas and suddenly the meeting turns very tense. Everyone has to look out for their own families, without depending on a group; it's just the responsible thing to do. But these are minority voices and eventually it's agreed the men will share the proceeds of any book or movie equally. Then they vote to make Illanes their official spokesman on the surface.

For Mario Sepúlveda, the appointment of Illanes as spokesman is a wound inflicted by men for whom he suffered and toiled. In this mine he's found a new calling, to be a voice for justice and truth and the workingman, but he won't be able to tell the full story of the miracle of Atacama when he reaches the surface. "They took me out of the leadership," he says. "It was the biggest betrayal I suffered in the seventy days I was down there."

Being trapped in the San José Mine has given Mario Sepúlveda a new sense of purpose, and it's made the greatness of his life of a workingman clear to him. He owes the mine something, and on the afternoon of October 12, with twelve hours to go before the first man ascends in the Fénix, he prepares to say goodbye to the mine's caverns and corridors. He goes to the spot where he slept in the Refuge and builds a memorial to his time there. When the last man leaves these passageways they will become a kind of time capsule, a landmark of Chilean and mining history destined to remain sealed and unseen for decades and centuries, perhaps. So Mario writes a letter to leave behind, and on the steel mesh that covers the stone wall near his bed, he places a piece of cardboard and writes down his full name and his date of birth, and the words "Mario Sepúlveda lived here from August 5 to October 13." He attaches some of the pictures he's been sent of his family, and places a plastic wreath around them, along with all the little Chilean flags he's collected. Then he goes about gathering rocks as souvenirs, from nearby on Level 90, the site of many adventures, including an attempt to set off an explosion to send a signal to the surface. "For me that level represented life, hope, a desire to live," Mario says. He's going to give the stones to André Sougarret, and to the engineer Andrés Aguilar, who played a key role in the rescue, and to the president, too.

As Mario does these things, Raúl Bustos prepares to leave, too. Raúl also gathers up rocks—but he takes them to the Pit and vents his fury by throwing them into that empty cavern. Take that, San José Mine! ¡Concha de su madre! Then, with other miners, Raúl takes some permanent markers and angrily scribbles graffiti on the vehicles that are the property of the San Esteban Mining Company, writing vulgar "thank-yous" to the mine owners and the owners' mothers. Raúl also says farewell to the space where he slept, up at Level 105. He's assembled a collage from the photocopied family pictures his family has sent to him, and he gathers these pictures that were his companions for so many lonely days when he longed to be home. "I looked at that, and remembered all the thoughts that went through my head there, thinking of meals I couldn't have, of birthdays that I missed," he says. He walks away to a quiet, private corridor in the mine where no one can see him, and then he sets fire to those pictures. "I wanted it all to go away, it was all bad memories," he says, and very soon the pictures become ashes carried away by the faint breeze in the corridor. When he returns to Level 105 he makes sure to erase any trace he might have left there. Like Mario Sepúlveda, Raúl Bustos imagines the San José in an unseen future, a time when men with cameras return to that place of Chilean history, those caverns where he suffered the deepest loneliness he's ever known. "I didn't want anyone else to see it, to come and say later, 'See, look, this is where Raúl Bustos slept.'"

"It was all very private, and it was mine."

Before he leaves the mine, Víctor Segovia takes a moment to write a final entry in his underground diary. "The earth is giving birth to its 33 children after having them inside her for two months and eight days," he writes. He takes time to reflect on his work history for the San Esteban Mining Company and its complex of mines, a series of memories that take him back to 1998, and all the jobs he had and the accidents that befell the men who worked alongside him during twelve years. He was a driller, an explosives handler, and a jumbo operator. A truck driver died in a cave in the adjacent San Antonio Mine, and another worker was killed in a road accident while on his way to the complex. He saw two men killed in the San José itself. Víctor doesn't blame the mine for their deaths, but rather "those who didn't invest enough money to make this a safe and secure mine." After all these years, the mine itself feels like a second home to him, he has an abiding and unbreakable affection for its improvised architecture, for the humility and crudeness of its serrated walls. Víctor draws a heart inside his diary, and writes "I LOVE SAN JOSÉ" inside it. The mine is like him: flawed and neglected but worthy of respect and love. "The San José was innocent," he writes. "The fault was in the people who didn't know how to run the mine."

Among the Chilean and NASA officers and engineers who shared ideas on the design of the Fénix capsule, there are two navy men, one American and one Chilean, with long experience in their respective country's submarine fleets. The submariners have both practiced rescues that are analogous to the mission about to be undertaken at the San José and they agree that the miners' ascent to the surface should follow a principle of submarine rescues: The strongest, most able-bodied man will enter the capsule first, because he'll best be able to deal with any complications that might arise. For that reason, the leaders of the Chilean rescue team decide that Florencio Avalos, the fit thirty-one-year-old foreman and second-in-command to Luis Urzúa, will be the first of the thirty-three men to ascend to the surface.

Before Florencio Avalos can come out, one man must go in. At 10:00 a.m. on the day the rescue is set to begin, Manuel González finds out that he will be the first rescuer to enter the Fénix. After a lunch in the beach town of Bahía Inglesa, a driver takes him to the San José Mine and the site of the Plan B hole in a van with tinted windows, which offers a kind of protection against the spectacle that's unfolding outside. The once-barren hillside around the mine is covered with people: family members and rescuers who can sense that a celebration is about to unfold, and reporters and cameramen who are there to broadcast that celebration to places near and far. After González goes down and the first miner comes up, two navy medics will follow him into the mine, but for now he is the focus of all the attention. After night falls and the final preparations of the capsule and the shaft are completed, González's tense face is broadcast on the giant screen down in Camp Esperanza set up for family members to follow the rescue. He looks up from the drill site and sees reporters and cameras lined up on a ridge above him, and many lights, as if he were entering an amphitheater and a great drama were about to begin.

González approaches the capsule and at 11:08 p.m. he's strapped inside. He's wearing a bright orange jumpsuit, a white helmet, and the expression of a man unable to completely suppress his fear of the unknown. He's been told that the system that will send him into the mine has been designed with several redundancies—among other things, the Austrian-built crane that will lower and raise the Fénix can lift 54 tons, or about one hundred times the weight of the capsule itself. And yet, when he places his feet on the steel floor of the capsule, he's standing over an open shaft that's as tall as a 130-story building; a free fall from top to bottom would last twelve seconds and result in certain death.

"Just stay calm," the head of the rescue team tells him. "I have total confidence in you."

The president and the minister of mining are there, too. "Good luck, Manolo," the president says, using a diminutive for his name.

At 11:17, the Fénix begins its descent. González can't see the shaft below him as the Fénix enters the mountain at an 82-degree angle. He has a radio, but the signal lasts only the first 100 meters or so, though there's a camera inside the capsule and he can communicate with hand signals if he gets into any trouble. "My mission was to make sure everything was working," he says, and he spends much of the seventeen-minute journey looking around. At about 200 meters down, he sees a trickle of water coming out of a crack in the shaft. The heat is building: It's a cool spring night on the surface, but in the deeper reaches of the mine it's a tropical summer. González feels a faint shift 150 meters from the bottom as the shaft bends from its 82-degree angle and heads straight down. His biggest worry is the state of the men when the capsule opens, the possibility that one of the miners will panic and try to force his way into the capsule before it's his turn.

In fact, down below, the miners are faced with the opposite problem. There's a man who doesn't want to leave the mine. An earlier test of the Fénix in the shaft has sent some stones tumbling down, and now Víctor Segovia is convinced that those loose rocks will cause the capsule to get stuck once he's in it. Even worse, the rumbling in the mountain has produced a huge crack in the wall of the cavern where the capsule will enter the mine. "I'm alive now, down here," he says. "Why should I go and die in that hole?"

"Look," Florencio Avalos tells him. "I'm going in first and if I get stuck none of us will come out." Florencio's words manage to calm Víctor. Later, as Florencio begins to prepare for his ride to the surface, the men bow their heads and say a final prayer. Sixty-nine days earlier they fell to their knees and asked God to lift them out of this place; now they ask God to protect Florencio in this first "journey" of deliverance. It's a final private moment before the public show of the rescue begins. Adding to the sense of theater is the bright light they've set up to illuminate the spot where the capsule will emerge, and also the camera that begins transmitting a live video image to the surface via the fiber-optic cable. The Chilean government, in turn, provides a feed to the assembled media and their cluster of satellite trucks, which send signals aimed at the desert stars, and several hundred million people around the world gather before screens large and small to gaze at the interior of the San José Mine as the capsule approaches. It's 11:30 p.m. in Santiago, before dawn in London and Paris, about midday in New Delhi, and dinnertime in Los Angeles.

The capsule slides down out of the shaft, and into the cavern at Level 135. Yonni Barrios, shirtless and wearing white shorts, is the first to move to the door and greet González, who steps out in his pristine orange jumpsuit. Yonni has tears in his eyes, González notes, and the two men quickly embrace. Turning to the rest of the men, the rescuer declares: "There's a shitload of people up there waiting for you guys!" As the rest of the men move toward him to shake his hand and embrace him, the rescuer makes a nervous joke: "You guys better not take advantage of me! Because there are two navy special-ops divers coming down after me and they're really good at fighting!"

To the men who have been trapped for nearly ten weeks, the tall González looks impossibly clean and fresh-faced. With a winning smile, big cherubic cheeks, and skin that's been colored by the days he's spent in the Atacama sun, he looks like a visitor from an impossibly bright and distant world. "We felt no other people existed," one of the miners says, and now a real, fully alive member of the human race is here among them.

To González, the thirty-three men look like primitives. Several of them are bare-chested and are wearing rolled-up shorts that look like "diapers" and cut-up boots, he says. "It was like they were a bunch of cavemen." González will be inside the mine for twenty-four hours, and later he will have a chance to explore a bit: Around the corner, he'll see a shrine to a man killed in an accident, and as he wanders more it's as if he's stepped back in time, to a simpler and more dangerous era of mining history. "They were completely without protection," he says of the men. He sees no respirators or safety glasses, and the heat and humidity are unlike anything he's felt in a mine before. The everyday working conditions are "inhuman," he says. One day in this mine would be a test of physical endurance, and yet the men here survived sixty-nine. How in God's name, he wonders, did they do it?

Now he must work to get them out. "I'm Manuel González, a rescuer from El Teniente mine," he says in a calm but authoritative voice. He tells them what the trip through the shaft will be like. "Look, you're just going to feel a little swaying, don't be afraid of it . . . The change in pressure will be noticeable." The final preparations include checking Florencio's blood pressure and pulse. "Ah, it doesn't matter," González says as he notes the high readings. "This is all for legal purposes anyway." He runs through a checklist and connects monitors to the harness Florencio is wearing and another to his finger. Less than fifteen minutes after González's arrival in the corridor at Level 135, Florencio Avalos is ready to step into the Fénix capsule. "We'll see each other up on top," he tells the other miners as he enters and González closes the door. A few seconds later the Fénix begins to rise, as smoothly and evenly as an elevator, and the capsule disappears into the shaft. As he rises, Florencio feels as if he were entering a body made of stone. "It feels good!" he yells down to the men below. *¡Se siente rico!* "It feels good in here!" The men yell back, their voices beneath his feet as he rises away from the caverns that were his home and his prison for ten weeks.

On the surface, Florencio's wife, Mónica, and his son wait for him near the opening: Mónica who once sleepwalked over this very mountain, and his seven-year-old son, Bayron. Farther down the hillside, in Camp Esperanza, María "the Mayor" Segovia watches the rescue on the giant television and thinks of those adult men squeezed into a stone channel and concludes: The mine is like a woman that's giving birth to them. Like many of the women who've been living on the property of the San Esteban Mining Company, she can feel the analogy inside her body. "If you're going to have a baby, you know that the baby might be born, but with complications, or the baby might not make it at all." A baby can be strangled by an umbilical cord, or he can get stuck in the birth canal and suffocate. As the men rise up through the stone, the capsule carrying them might fall back down into the mine, or the mountain might rumble again and destroy the borehole, causing it to crack and trap the Fénix and its passenger inside. María Segovia has given birth to four children, and now the men whose lives she's been fighting for will rise up through a 2,100-foot birth canal carved into the mother mountain. If the Earth doesn't want to let them go, they won't be able to leave, she thinks. But maybe the Earth doesn't want to hold on to them any longer.

Inside the capsule Florencio is awake for this birth, watching as a small light illuminates walls of carved stone that pass before his eyes. He's wearing the same faded red helmet he put on when he entered the mine on August 5. It's a few minutes before midnight and on his slow journey upward, October 12 becomes October 13. He can hear only the rattling of the capsule: It sounds as if he were riding an old roller coaster. He feels the swaying back and forth, but Florencio stays calm for the thirty-minute journey, because his long ordeal inside the mountain is nearly over. He is alone, but on the surface an audience of 1.2 billion people is waiting for him, their eyes focused on a cylinder jutting out of the mountain.

Florencio begins to remember the events that unfolded inside this mountain and then other memories come, from his life outside: the day he met the woman who would become the mother of his children, the days those boys were born, the days his sons headed off to school. He's had a good life, he realizes, and today he's been blessed again, rising from the stone caverns of the mine in a capsule, being pulled up by men and women he cannot see. He feels the air turn thinner and lighter. His ears plug up, and then they pop. A breeze from the surface flows into the capsule as it enters the final section of the shaft, and for a few moments he is surrounded by steel walls and the rattling sound disappears, replaced by an eerie quiet. The radio squawks to life, and he hears people, the shouts of men calling out instructions to him and to one another, voices on the surface floating above his head. There is a sudden burst of applause. With the Fénix still slowly rising, light and color flood in from the outside, and Florencio looks up to see a sunburned man in a white helmet peering at him through the steel mesh of the capsule door.

PART III

THE SOUTHERN CROSS

18

IN A BETTER COUNTRY

On the night of August 5, Bayron Avalos pronounced his father, Florencio, dead, but on October 13, he sees him resurrected from the rumbling mountain just a few minutes after midnight. With the world's cameras trained on him, Bayron breaks into tears and begins to bawl uncontrollably. The First Lady of Chile, who is standing alongside him, tries to comfort him. Florencio Avalos emerges from the Fénix capsule and falls into a silent embrace with his wife and son, and clasps the hands of the president, Minister Golborne, and the other leaders of the rescue team. While hundreds of reporters and anchors from around the world watching the events comment on the rescue unfolding before their eyes with breathless enthusiasm, the scene around the capsule itself is subdued and sadly quiet—especially compared with what will come about one hour later, when Mario Sepúlveda rises toward the surface. The man with the heart of a dog can be heard yelling when he is still twenty meters from the top. "*¡Vamos!*" he cries, with a disembodied caveman's shout that spills forth from the top of the cylinder jutting out from the shaft. "*¡Vamos!*" he shouts again, causing the rescuers and his wife to laugh, and when the Fénix finally rises up out of the shaft, he gives an animal screech that causes everyone to laugh more. The door opens and Mario quickly embraces his wife, then reaches down into a bag he's brought from the interior of the mine. He passes out slate-colored rocks from Level 90 as souvenirs to Minister Golborne and the leaders of the rescue team and to President Piñera, and he takes off his helmet, like a gentleman or knight errant, and bows his head to greet the First Lady. Moments later he's embracing a group of rescue workers with shouts of *"¡Huevón!"* and then he leads everyone present in a "Mineros de Chile!" chant, raising his arms with liberated, frenetic energy until one of the rescue workers finally stops him and tells him to take off his harness, please—there are, after all, other guys down below waiting to be rescued. Finally Mario is lowered onto a stretcher and carried away to a nearby triage room, and will later be flown (as will all his other colleagues) to a hospital in Copiapó.

Juan Illanes is the third man out, followed by Carlos Mamani, who is greeted at the top of the shaft by the president of Chile, and later at the hospital by President Evo Morales of Bolivia. He is followed by the teenager Jimmy Sánchez, who is met by his father, and then Osman Araya, who led so many prayer sessions, and José Ojeda, who crafted the famous note about "Los 33." Claudio Yáñez, who looked so weak he was like a newly born colt, emerges with his chiseled young features in the soft, muted light of an overcast morning, and clasps his girlfriend and the mother of his children in a rocking embrace. He's followed by Mario Gómez, the truck driver who went back down to make a few more pesos from an extra load and who now insists on taking a few moments to fall to his knees in prayer before the capsule. When the tenth miner comes out, the sun has burned through the morning overcast, and the men and women at the top of the shaft call down to Alex Vega: "Put your sunglasses on!" Alex once wondered if the darkness in the mine would make him blind, but now it's the potential damage caused by desert sunshine he has to worry about, and he has his sunglasses dutifully on as the capsule emerges. His wife, Jessica, who refused to kiss him goodbye on the morning of August 5, gives him a kiss and a hug of cinematic passion and length, and filled with so much longing and sorrow that all the people around them stop applauding and the sound of Jessica's tears can be heard by everyone present.

"Don't cry," Alex says. *No llorís.* "It's over now."

Jorge Galleguillos ascends next, followed by Edison Peña, who comes out saying: "Thank you for believing we were alive. Thank you for believing we were alive." He repeats these words a few times more, even as he falls into the arms of the woman who's asked him to marry her. (He will not.) A short time later word will come from Memphis that Edison has been invited to Graceland. Carlos Barrios, who joined a pair of expeditions seeking a way out of the mine, is greeted by his wailing father: "*Tranquilo, ya*," Carlos says. Next comes Víctor Zamora, the poet. The psychologist, Iturra, has made good on his promise to deputize Zamora's son Arturo as a "junior rescuer," and the boy is wearing a white helmet with a Carabinero police symbol on the front, and unlike the other children at the site, he's allowed to walk up to the capsule itself and help open it. Just after noon Víctor Segovia emerges, the precious diary still in his possession, followed by the tall, quiet truck driver Daniel Herrera, and Omar Reygadas, the man who took a flame to the bottom of the mine. The eighteenth and nineteenth men out are the cousins Esteban Rojas and Pablo "the Cat" Rojas, who was in the mine doing makeup work because he had missed time for his father's funeral. It's midafternoon by the time Darío Segovia emerges. His sister María, the Mayor, is not present (she's remained in Camp Esperanza below), and he's greeted instead by his partner, Jessica Chilla. She holds his face, and touches his limbs, something she will do again and again when she sees him at the hospital, like a mother checking on the health of her newborn baby. "I wanted to see if he was whole," she says. "I don't think he was aware of what was happening, he was so nervous." The magical and surreal sense that she's witnessed the birthing of a middle-aged man colors those first hours Darío is back with her. "It was like he was starting a life he never thought he would live."

Yonni Barrios hears cries of "Doctor!" when he reaches the top. He doesn't see his girlfriend, Susana Valenzuela, right away, but she's there, standing next to the minister of mining, following several days of private family drama, a few elements of which have played out in the world media. Yonni told Susana in their final videoconference not to worry, that he would make sure she was there to greet him on the surface. He will be like Tarzan, king of the jungle, he said: He will simply speak and all the animals in the jungle will work his will. And so it is. The rescue team tells Yonni's legal, estranged wife, Marta Salinas, that Yonni wants his live-in girlfriend to have the honor of greeting him. Marta is left to make statements to the press lamenting her husband's choice. "She's welcome to him . . . I'm happy for him." On the morning of Yonni's rescue, Susana finishes her work at the camp kitchen frying up fish, changes into fresh clothes, and gets a police escort up to the rescue site.

Susana watches Yonni step out of the capsule, noticeably thinner. He's taking off his harness, and has his back to her when she calls out to him, in a low voice: "Hey, Tarzan."

Samuel "CD" Avalos is the twenty-second miner out, followed by the young Carlos Bugueño, and then José Henríquez, the Pastor, who is greeted by his wife of thirty-three years, Blanca Hettiz Berríos. It's late afternoon and the rescue site is in shadows when Florencio's brother Renán Avalos reaches the top next, followed by Claudio Acuña, who emerges to the sound of his infant girl crying loudly. Next is Franklin Lobos, the soccer player, who hugs the twenty-five-year-old daughter whose tears helped bring the federal government into the rescue. She gives him a soccer ball signed by his family members, and Franklin takes a moment to juggle it with his feet. Night falls over the site as the final miners come out: Richard Villarroel, who will not leave his soon-to-be-born son to grow up fatherless, as he did; Juan Carlos Aguilar, the head of the mechanics crew; Raúl Bustos, the man from Talcahuano; and then Pedro Cortez, who notes how sweet and fresh the air tastes, up here where people are meant to live. The thirty-second man out is Ariel Ticona, who will have to wait just a short while longer before meeting his baby daughter for the first time. Luis Urzúa is the last man of the A shift to leave. On August 5, he went deeper into the crumbling mine to make sure all the men in his shift were accounted for, and tonight he's completed an everyday supervisor's ritual: He's left his work site only after all the men in his shift are out of the mine and accounted for. His arrival at the surface is greeted by horns and sirens that echo across the mountain. Urzúa is embraced by his son, and then by the president, and he begins to speak to Piñera in a low, exhausted voice that the nearby microphones struggle to capture. "As the *jefe*, I hand over the shift to you," he tells the president. "Like a good *jefe*," the president says. With those words, the public ordeal of the thirty-three men of the A shift of the San José Mine comes to an end.

•

The rescuer Manuel González is the last man left inside the mine. When the Fénix capsule descends to take him out, he faces the remote camera that's broadcasting to the surface and takes a bow. González enters the capsule and rides up to the top, where he is greeted by President Piñera. The president helps roll a steel cover over the top of the Plan B shaft and delivers a speech in which he praises the courage and tenacity of the miners and their rescuers. "Today Chile is not the same country it was sixty-nine days ago," he says. "The miners are not the same men who were trapped on August fifth. They have come out stronger and have taught us a lesson . . . Chile today is more united and stronger than ever." President Piñera will never again be as popular as he is at that moment.

At Camp Esperanza, the several thousand journalists, rescuers, and family members begin to pack up and leave. The tents, the small school, the kitchens, the altars, and even the flags disappear. The mine is already taking on the desolate, lonely appearance of a desert archaeological site as María "the Mayor" and the other siblings of the rescued miner Darío Segovia putter about and make sure everything is cleaned up and in order. "We were the last people to go, there was nobody there," María says. Only some police officers remain, as guards of the empty property. At four in the afternoon on Day 74, "We closed the camp, as a family." Darío is in the hospital in Copiapó, and then he's busy with his immediate family, and María decides she'll allow him the space he needs to get his life in order, and she takes the bus back to Antofagasta without ever seeing the brother she worked for ten weeks to free. "I left the camp happy, because we had won, we had won his life, but I was sad because I couldn't see him. That marked me." In the weeks to come, María returns to selling pastries on the beach from a cart, under the hot sun, and at home she watches on television as Darío becomes a "magnate," traveling the world and accepting honors. They speak on the telephone, but a year passes without the two siblings seeing each other. One day, however, María receives a letter from her brother in the mail. "I'm very proud of you, Madame Mayor," it begins.

19

THE TALLEST TOWER

On October 16, at a meeting hall of the Chilean social security administration in Copiapó, Juan Illanes leads six of his fellow miners in their first official press conference. The men sit behind a cluster of microphones, their skin still a sickly gray after ten weeks underground. Thirty-two of the miners have been released from the hospital—only Víctor Zamora, suffering from rotting teeth, remains under medical supervision—and clusters of reporters have shown up at their homes. Mario Sepúlveda was escorted away from the hospital in secret, his head covered with a blanket to avoid the media throngs seeking to speak to him. Now Illanes asks the press to respect their privacy. "Leave us enough room so that we can learn how to deal with you all," he says. He asks the media to refrain from trying to "destroy the image" of the miners as a group, and especially of men like Yonni Barrios, who has become the target of many mean-spirited stories, "complete with nicknames," that make fun of his amorous entanglements. "Please consider his emotional state of mind," Illanes says. In Latin America, as elsewhere, the media builds up heroes and then takes delight in destroying them, especially when they choose not to cooperate with the machinery of celebrity, and Illanes can feel how quickly this pack of questioners might turn on him. He answers some surprisingly skeptical questions about why anyone would want to work in such a dangerous mine in the first place—"I needed the money," he says. But he declines to talk about how the men survived for the seventeen days before they made contact with the outside world. The men have a pact of silence, and an agreement to share in the proceeds of a book and movie, and they won't be talking about those seventeen days, Illanes says. In the questions the reporters ask at this press conference and while staking out the miners' homes there is a suggestion of the sublime and ridiculous stories that the media imagines must be waiting, unspoken, on the lips of these pale men. Did you fight among yourselves? Did you ever think about sex? Did you see the face of God? Did you consider cannibalism? Already, the Chilean media is hinting that all might not be what it seems among the heroes of the San José Mine. Clearly the men were divided, and some outlets have reported what a rescuer overheard a group of men say when Mario Sepúlveda was leaving the mine in the Fénix capsule: "We're lucky we're getting rid of that guy!"

Reporters are surrounding Víctor Segovia's house on Chalcopyrite Street in the Los Minerales neighborhood, and when he passes through one media phalanx to get to his front door, he finds more reporters who've talked their way inside, including one of the more famous media personalities in Chile, Santiago Pavlovic, the eye-patch-wearing host of the show *Informe Especial*. There's another reporter in his kitchen talking to Víctor's mother, and a reporter from Asia. Víctor wants to go to his backyard and have a beer, but there's another reporter there, too. His relatives are telling him: "Please talk to these reporters so that they leave." While all of this is happening, Víctor is trying to comfort his seventy-seven-year-old father. The elder Segovia is ill and losing his memory, and when he casts eyes on his miner son for the first time in ten weeks he begins to weep. "Never, never had I seen him cry before," Víctor says. "He was always a hard man." Above all, Víctor is confused by the way everyone is treating him, here in his own home, as if he were a celebrity, as if he were rich, with a mixture of awe and resentment. They are impatient with him, they want to see him smile, they want to hear what his plans are now that he doesn't have to work anymore, because everyone in Chile knows that one of the country's richest men said Víctor and the other thirty-two miners would be millionaires. Even Víctor's ex-wife, who dumped him years ago, suddenly wants to make amends and is asking for forgiveness, and it's all as strange and dreamlike as that bearded man with the eye patch, somehow transported from the television into Víctor's living room, staring at him and asking, "Can we talk?"

The media loves the thirty-three—and the media is starting to resent them. Chile's newest national heroes are a bunch of ordinary working stiffs who have the temerity to ignore the media's most pressing questions, because they've got plans to make storytelling money all on their own—not in Santiago, but rather in Hollywood and New York. A few sell small parts of their story for sums big and small—"he charged us fifty dollars but it felt like he was holding something back," says a Japanese reporter after visiting one of the men. If the Chilean media won't be allowed to make them heroes, they can very easily tear them down and make them to objects of populist scorn. For starters, several media outlets begin to point out, there's the price the country paid to rescue them: at least $20 million, the government estimates, including airfare for technicians brought to the site, $69,000 on the Fénix capsules built by the navy, and close to $1 million spent by the national oil company on fuel for various drilling machines and trucks. On October 19, the tabloid La Segunda runs a story that adds up the cost of all the gifts the miners have received: more than $38,000 each (or about 19 million pesos) in "vacations, clothing, and donations," the paper reports, including Oakley sunglasses worth $400 each and the newest version of the iPod touch, donated by Apple, and planned trips to Britain, Jamaica, the Dominican Republic, Spain, Israel, and Greece (the miners having been invited to visit those places by assorted officials and entrepreneurs). In the end, not all those trips will come off, and only a handful of miners will travel on most of them. But Luis Urzúa can sense a shift in attitudes. "After that story in La Segunda, people started to think we were getting rich. They looked at us differently." In the short term, however, people are falling over themselves to give the men gifts. A few days after La Segunda's tally comes out, Kawasaki Chile announces it's giving a new motorcycle to each of the thirty-three men. This is our most expensive model, says the general manager (they cost 3.9 million pesos each). "Above all, the miners deserve it," the executive tells a television reporter, while also managing to link the Kawasaki brand to the miners:

"These men represent hard work, sacrifice, tenacity, the ability to over-come obstacles—qualities that are also represented by Kawasaki, one of the most important companies in Japan." Franklin Lobos accepts the gift on behalf of his colleagues, and says something the men have repeated again and again since coming up from the surface. "We are not heroes, like people say. We're just victims. We're not movie stars, or Hollywood stars."

A few days later, Ariel Ticona finds himself in a Madrid studio with his wife and new daughter, Esperanza, answering questions from a talk-show host for a Spanish television show. We have a gift for you, the host says, and on cue a young woman in a form-hugging dress emerges from offstage, pushing a brand-new stroller. Ariel's next stop in Spain is the Santiago Bernabéu stadium, that temple of world sport that the Real Madrid soccer team calls home. Along with three other miners, Ariel gets a VIP tour that includes a walk on the field itself—with a television camera in tow. "This is the most beautiful thing I've begun to experi-ence," Ariel says as he stares up at eighty-five thousand empty seats through his Oakley sunglasses. There's something magically innocent about the way Ariel smiles and his face widens as he turns to take it all in.

In the first weeks after emerging from the dark corridors of the San José Mine, the thirty-three survivors are standing in an arena of public adulation, while also living with the private memory of their humble backgrounds and the ten weeks they spent at the mountain's mercy. Edison Peña is soaking up as much media attention and praise as any-one—a man who jogged and sang "Heartbreak Hotel" underground, after all, seems to represent the epitome of the strength and joyfulness of the human spirit. But after surviving inside a thundering mine, Edison can see there is something cruel about being on the surface, watching people go about their "normal" lives. Edison's mind has lagged behind his body: It's still in that mountain that's falling on top of him over and over again; it's still trapped behind the stone guillotine. The mountain stays with him as he travels the world as an ambassador of Chile, and of mining and jogging culture, visiting Tokyo and Tupelo, Mississippi, and many other places in between. It especially haunts him back home in Santiago. "All the evenness of life, the 'light' part of it, really stunned me," Edison says. "It shocked me to see people walking around, living normally. It shocked me because I would say 'Hey, where I come from isn't like that. I come from a place where we were fighting desperately to live.' I came out to life and I found this shit called peace. It threw me off. It threw a lot of us off." In the mine Edison ran to forget where he was, and now on the surface he runs to forget where he is. On October 24, eleven days after the rescue, Edison Peña participates in one leg of a triathlon in Santiago, running 10.5 kilometers. "The doctors, the psychologists, they have me on a strict regimen I have to follow," he tells a television reporter at the race. "I feel sort of abnormal." Edison confesses to other people, privately and publicly, to feeling unstable, but that doesn't stop him from accepting an invitation to watch the New York City Marathon. In New York, he sings an Elvis tune on the David Letterman show, and answers questions at a press conference before the race. Why did he run in the mine, someone asks. "I was saying to the mine, 'I can outrun you, I am going to beat back destiny,'" Edison answers. Edison was more of a cyclist before the accident than a jogger, but he decides he'll not just be a spectator at the New York City Marathon, he'll run it, too, and try to finish. A doctor and members of the local running club that invited him tell him that entering a marathon without having trained for it is a foolhardy thing to do, but Edison is determined. Sure enough, his knees start to give out after an hour or so, and he ends up walking about ten miles of the race, but he finishes (with a time of 5 hours, 40 minutes, and 51 seconds), thanks in part to two Mexican immigrant restaurant workers and running-club members who escort him along the entire route. "In this marathon I struggled," Edison tells the press afterward. "I struggled with myself. And I struggled with my pain." Several of his colleagues in the mine will say afterward that New York was bad for Edison Peña: It was there he started to fall deeper into alcohol addiction. "If we had really been united, as thirty-three men, we would have looked after Edison and he wouldn't have had the collapse that he did," says the young miner Pedro Cortez. In New York, Edison starts to get in arguments with his girlfriend over whether he should be traveling, but he can't say no when he's offered another trip. "You start to become a puppet. We became puppets. We're going here, we're going there. 'Stand like this. Over here, over there, under the lights.' We wanted to go out and bite the world. We had been born again and aaahhh . . . That first year . . . I wouldn't know how to explain it, but it was rough. Pack your luggage, stand in line. Do this, do that. Do! I think, looking at it honestly—it's like we lost our lives." Edison is falling deeper into the spiral when he reaches Memphis and Graceland in January, just in time for Elvis's birthday. At another press conference, Edison sings a few lines from "The Wonder of You," in an accented but swooning baritone. "When everything I do is wrong / You give me hope and consolation." He gives his Graceland audience the heartfelt rendition of a man who's living inside the bluesy world of the song, and several of his listeners, who have no way of knowing what's tormenting Edison, scream with approval even before he finishes.

For the first few weeks, the miners talk to psychologists and therapists. "My girlfriend says I wake up yelling in the middle of the night," Carlos Bugueño tells a psychologist, and he later gets pills that help him sleep. "At night, all the memories come back," Pedro Cortez says. During the day, long silences are haunting, and so are loud, cracking noises. Pedro gets into a van with a group of five mine survivors, and he and the others all fall to the floor at the sound of a motorcycle's backfire. Bugueño and Cortez join a large number of the survivors at a clinic of the Chilean social security administration in Copiapó for a group therapy session. When the therapist closes the door, for privacy, several of the men stand up to leave. "You have us locked in here," one says. "Don't shut us in!" The sight of the closed door, combined with the familiar faces of the men with whom they were trapped, sends several of the men back to their underground emotional state. "We went to the windows and opened all of them," Cortez says. "They couldn't have a meeting with all of us together, because the stress was just too much." They will continue with individual visits to therapists, but most will last only a few weeks.

The psychologist, Iturra, is sharply critical of the postrescue treatment afforded to the men. His recommendations are largely ignored, including his argument that the men should return to a new, moderate work routine (aboveground, of course) after a week or two of vacation. Instead, most of the men continue to sit at home expecting to enjoy the fruits of their worldwide celebrity, and feel obliged to attend all the official and unofficial events in their honor. "They became trophies," Iturra says. "They became symbols." If you make a man a symbol of things that are bigger than any one person can possibly be, you risk stripping that man of his sense of who he really is. "The worst thing that anyone did was to call them heroes," says one of the miner's wives. The same government that worked so hard to pull off a technical feat never before seen in history, to rescue thirty-three ordinary men, should have realized that those ordinary men were about to undertake an emotional journey that was also without precedent. But their surface suffering unfolded, for the most part, in the private world of each man's home, and no official stepped forward to boldly take charge of their recovery.

Instead, in late October, all thirty-three men are asked to visit La Moneda presidential palace for a public celebration. For a handful, like Florencio Avalos, it will be one of the last times they submit themselves to display. "I went to La Moneda because I had never been there and I always wanted to see that. After that, I never went to anything." The thirty-three miners receive Chile's recently minted Bicentennial Medal for heroic actions that encapsulate the proudest qualities of the two-hundred-year-old republic. They listen to the president make a speech in which the rescue becomes a metaphor for what Piñera hopes to accomplish during the rest of his still-young tenure. The president talks about building a country without poverty, a society that treats its workers better, and he says that the 700-meter rescue shaft crafted "by our engineers and technicians," and which served as "a bridge of life, faith, hope, and liberty," won't be the last great project Chile undertakes.

For the thirty-three men of the San José Mine, as for Chile itself, the future appears to be filled with promise. The miners are rising to new heights, literally, when they visit the offices of their new Santiago lawyers in December. Following a series of recommendations, the miners have chosen the biggest law firm in the country, Carey and Company, to transform the oral agreement they reached underground into a legally binding document. Carey and its specialists will also represent them in the negotiations for movie and book rights, contacting talent agencies in New York and Los Angeles, and the Washington, D.C., law firm Arendt Fox. But first the men must see exactly what Carey has to offer, which requires a group visit to the firm's new offices on the forty-third floor of what is, at this moment, the tallest building in Chile—the recently completed Titanium tower in a swank neighborhood of Santiago known as "Sanhattan."

Carey has assigned ten lawyers to draft the agreements. They are among the country's best attorneys, bright and ambitious multilingual men and women educated and trained in law schools and law firms in Chile, Europe, and the United States, but when they finally meet the men of the San José Mine, they are momentarily awestruck. "When you looked at them, you felt this overwhelming feeling of patriotism," one of the lawyers says. Looking at these Chilean everymen is like looking at the flag, or the Andes, though the feeling of awe dissipates fairly quickly as the lawyers get down to business. They've prepared a twenty-page contract, and a PowerPoint presentation, but the interest of the miners in the subtleties of intellectual property law as practiced in Chile and the United States soon wanes—one of the lawyers notes a young miner in the back playing a game on his phone, looking childlike.

When the presentation is done, the lawyers leave the miners alone in a meeting room to decide whether they'll agree to it. The corporate meeting room on the forty-third floor is one of the most impressive in Latin America—it's called the Manquehue Room because it faces a peak in the Andes of the same name. The discussion is short and civil, even though several of the miners are furious at Mario Sepúlveda because he granted an interview to a BBC journalist who is writing a book from which the miners won't make a cent. Mario has also spoken independently with a Latin American movie producer who said he could offer an astronomical sum of money for movie rights. The agreement with the Carey firm will create a new entity called Propiedad Intelectual Minera, S.A. (Miner Intellectual Property, Inc.), but that agreement is worthless unless all thirty-three men agree to join it, including Mario Sepúlveda. The man with the heart of a dog is, at that moment, one of the most popular men in Chile, and he could easily go off and sign his own deal. "Mario was very much aware of his power," says one of the lawyers. Mario brags to the lawyer that he can make a call "and have tea with the president this afternoon."

Mario Sepúlveda, feeling the pressure of his colleagues, and faced with the prospect that they might all lambast him in the press if he does otherwise, agrees to sign. Propiedad Intelectual Minera is born and the men, who on August 5 entered a subterranean workplace that was among the least desirable places to work in Chile, can take a moment to feel like corporate bigwigs in the tallest building in Chile. The skyscraper isn't even a third as tall as the San José Mine is deep, but the view through the meeting room's windows is limited only by the smog of Santiago's southern hemisphere summer. The building is flooded with light, and from their perch in the sky the former miners can see a new highway being built beneath the Mapocho River, a massive construction project that is a symbol of Chile's entry into the "first world." A new Chile is being born, and its future is boundless, and so is theirs. They are symbols of that nation's fortitude, the president himself has said this, and Congress has given each of them a medal forged from ore extracted from the mountains of Chile by men who suffer and labor deep below the ground.

20

UNDERGROUND

In their widely publicized visit to Disney World in January, the men of the San José Mine wear yellow faux mining helmets with black mouse ears attached. In February, twenty-five of the miners visit the Holy Land, and the Israeli ministry of tourism gives each man a hat emblazoned with the slogan "Israel Loves You." The men of the San José are grateful to the Disney Corporation for the opportunity to take their family members to "the Happiest Place on Earth"; and they're grateful to the government of Israel for the chance to visit the Church of the Holy Sepulchre and the Jordan River and so many other sacred places and thus pay homage to the faith that helped save their lives. But mixed in with that gratitude is the oddness of the celebrity treatment that follows them as they circle the globe. "To be treated like a rock star—that was stressful," says Pedro Cortez, who goes on both trips. "We got to Disney World and people wanted to touch us. As if we were God, almost." A visitor to the Magic Kingdom sees a man in a yellow helmet walking down its pretend Main Street and is told he's "one of those Chilean miners." The visitor remembers the story attached to that helmeted man: He's been resurrected from the deepest stone tomb in human history. How often is one in the presence of a miracle? So they point cameras in the helmeted man's direction, follow him a bit just to watch him walk. "Yes, it's a miracle we're alive. We're grateful to God and all the people who helped us," Pedro says. "But it was like being in a movie about Holy Week, where Jesus is walking and everyone is following him." This odd behavior of strangers continues in the Holy Land itself.

When he returns to Chile, Pedro decides it's time to stop feeling like a hero or a character in a Bible story. He's going to get his life back in order. For starters, instead of buying the pricey yellow Camaro he dreamed of when he was trapped in the mine, he buys a used Jeep. More important, he decides to enroll in a university, to get a degree in electronics. But as he begins attending classes, there is the small problem of being the only worldwide celebrity on campus. Journalists stake out his classroom to talk. "I wanted to be relaxed but everything turned against me," he says. Television news reports and long silences both trigger memories of the mine, and the faces of his girlfriend and relatives trigger feelings of inadequacy. One day Pedro leaves class, weeping, and misses two days of school. "I felt like I was drowning." He believes he is disappointing all the people around him, that he'll never live up to their expectations, but he struggles to explain these feelings to the professional who is supposed to be helping him. "Even the psychologist didn't understand," he says.

Víctor Segovia does not suffer from nightmares or a sense of worthlessness in the first months after he's liberated from the San José Mine. But his cell phone is ringing constantly with the voices of friends and relatives, who see him as the person who can summon the magic that will solve their problems: not because he's a living miracle with access to the divine but rather for the money that's in his pocket. They call with a series of laments and requests: "Víctor, I don't feel well." "I've got a problem at home, *huevón*." "I need a million pesos" (about $2,000). "They're going to repossess my television, my dining-room furniture, help me!" Víctor says his friends and relatives are treating him "like a bank." "I'd have a guy call me and ask for forty or fifty large," he says, or about $80 or $100, "and he wouldn't even invite me to a beer first." "The whole thing was just to get money out of you." He is surrounded by many concentric circles of need: eventually friends of relatives, and friends of friends start to ask him for help. When Víctor finally stops loaning money, he realizes he's doled out about 6 million Chilean pesos (about $12,000, roughly a year's salary), most of which will never be paid back.

I start to meet the miners at about the same time it's beginning to dawn on them that their postrescue bonanza isn't going to be as big, or last as long, as they expected. Like Víctor Segovia, they begin to burn through their Farkas money fairly quickly—a million Chilean pesos doesn't go as far as it used to. And with much of Chile believing they'll get rich selling their story, no one follows through on Farkas's call to raise one million dollars for each of them. Richard Villarroel is the very first miner I talk to in private, at a table in an empty restaurant in Copiapó. He tells me about hitting the drill bit with a wrench when it came through, about growing up without a father, and about the recent birth of his first son. He brings the conversation to the present and his mental state, because now that he's no longer visiting foreign lands and he's home, the burden of what he's been through is all the more apparent. "I'm in the hardest part right now," Richard tells me. "I don't have any feelings. I'm a more serious person. A harder person. I don't cry for anything. My wife has noticed it, too. Whatever happens around me, it's like I don't care. I have this disorder in my head. I could be talking to you one moment, and then suddenly I lose the thread [*se me va la onda*]. I have to wait for you to bring up what we were talking about so that I can remember what it was." I ask him if he's been seeing a psychologist or a psychiatrist. I was, he says. But the professional treating him said: "You're fine. You can go." To which Richard responded: "I am? But I don't feel the same. Talk to my wife. She'll explain to you how I was before and how I am now."

As I meet the miners and travel to their homes, several of their wives and girlfriends express that same thought: The man who came out of the mine isn't the same one who went in. "The Arturo we used to have here in the house stayed behind in the mine," Jessica Chilla tells me, using her partner Darío Segovia's middle name, the one they always use at home. The new Darío Arturo Segovia is stoic and emotionless. "You can punch him, and he won't say anything. He doesn't feel anything." Even his six-year-old daughter says, "He isn't the same Arturo." Jessica longs to return to their old, soothing daily routines, the simple pleasures of taking turns picking up their daughter from school.

"We had a whole system of life," Jessica tells me in their living room.

"Yes, a system of life that was beautiful," Darío says.

"He even cooked for me," Jessica says.

"I cooked," Darío says.

"Now he doesn't cook," Jessica says, and she laughs, because it really was remarkable for a tough miner like Darío to make meals for her, and to do it as lovingly as he did. She also laughs because as much as Darío Arturo has changed, she can sense, by this point, that he's starting to get better. "Two or three months ago, he was much worse."

Over at the home she shares with Yonni Barrios, Susana Valenzuela has witnessed the suffering of her "Tarzan." When the sun goes down each day and the windows turn dark, he becomes depressed. Yonni wakes up in the middle of the night sometimes and puts on his old helmet, and sits in the living room in the dark with the mining lamp on, as if he were back inside the caverns of the San José, listening to the distant thunder. Sometimes he begins to scream and pounds at the cushions of their sofa. "I didn't know what to do," Susana says. This goes on for several nights, until finally Susana turns on the living-room lights, grabs him, embraces him, and says, "Wake up, wake up, *huevón*, it's over already." Later, he sleeps all day, all night, and he sleeps and sleeps so much it can't be normal. Then he can't sleep at all again, and Susana makes him tea and milk, and brings it to Yonni on a tray and pretends it's his birthday, singing to him like a little kid. "*Estas son la mañanitas . . .*" She does this for a few days, each day another "birthday" of tea and warm milk and singing, and she has him go back to his psychiatrist, and after a while he starts to calm down a little.

At about this time, I show up at Yonni and Susana's house for the first time, and I talk to Yonni for more than two hours in a living room dominated by several photographs of Yonni and Susana embracing in the days after the rescue, Yonni looking lean and pale and exceedingly happy to be in his girlfriend's arms. When he remembers the collapse and the days of starvation, he sheds many tears, but it's clearly cathartic for him to finally share the story with someone who will tell it to others. "I liked when you came and he talked to you, because it's like he let go of it," Susana tells me later. "He was going to be loyal to his promise, and he wasn't going to talk to anyone else."

The second time I go to Yonni's home, several months later, it's after his wife, Marta, has sued him. I arrive with the screenwriter José Rivera and the film producer Edward McGurn, and we ask Susana about her boyfriend's wife, and she suggests that we go talk to her. "Marta lives on the next block," Susana says. "Yonni can show you where. Yonni, go show them," she commands. When Yonni looks reticent to do so, Susana calls out with a hearty laugh: "Don't worry, I'm not going to hit you!"

Yonni walks to the end of his block and crosses the street. The man famous the world over as the Don Juan of the San José Mine points at a house a few doors away, with a faint smile that's either meek or devious, I can't tell which.

We talk to Marta Salinas for a few minutes, as she stands behind the candies and sundries lined up under the front counter of her neighborhood store, the one Yonni took out a loan to pay for. Marta says she took the letters Yonni wrote to her from inside the mine and sold them to an American journalist. When we're finished talking, she asks, "Did Yonni get the money from the movie yet?"

"No, señora," I answer. "Not yet."

Few people besides the thirty-three miners know precisely what José Henríquez said when he was underground, but around the world he's earned the name the Pastor. A few weeks after he emerges from the San José Mine, Henríquez gives a talk to a rousing crowd of believers at an auditorium-size Evangelical church in Santiago, with several of his fellow miners present. "I could see before me, thirty-two men humbled before God," he says from the dais, briefly describing the prayers he led underground. "Now, I thank the Lord for this opportunity to testify to the great power of God. What God did in that place is undeniable. And let no one rob God of that glory. That's why we're here," he says, and he raises his fist defiantly, like a man who's won a great victory for his cause. In the days and weeks to come, José Henríquez could very easily transform his fame as El Pastor into a lucrative speaking tour, because the agreement the thirty-three men have signed allows each man to give talks as long as they don't reveal the essential secrets about their first seventeen days. But Henríquez stays home, mostly, and downplays his role. When he does speak, he takes pains to say that he is not really a pastor. "I think what God saw in the mine and what convinced God was humility," Henríquez says in an interview with a Christian broadcaster. Humility requires Henríquez to recognize he is not a pastor, because men who have that vocation suffer to bring the word of God to others, as his grandfather did, bicycling for many years from one place of worship to the next. "I'm just a man who went into a mine knowing what the consequences might be."

Florencio Avalos, the foreman who was the first man out of the mine, turns down all the trips to which he is invited, including one to Great Britain at which his presence was specifically requested. On the anniversary of the accident, there's the dedication of a new exhibit about the miners at the Atacama Regional Museum in Copiapó, but he doesn't go, even though the president is speaking and has requested Florencio's presence, and the location is just a ten-minute drive from his home. "Those things don't interest me," he tells me. I visit his home three times to listen to Florencio recount his experiences in the San José Mine. He speaks with a sense of wonder and gratitude, though he doesn't appear to suffer with these memories as much as his colleagues do. Florencio has settled back into a routine, taking an aboveground job with a mining company, while his sons go to school. "I work so that my sons can study," he says. "If I don't work, they can't go to school." We sit and talk in his living room, in his two-story condo-style home in a middle-class Copiapó neighborhood. He invites me to sit down and have lunch, at the same dining room table where his wife prepared soup for him on August 5. Later his teenage son, César "Ale," leaves for school, and I watch as Ale stops to give both his mother and father a kiss on the cheek goodbye. North Americans don't often see teenagers treat their parents with such affection, and even though it's a common gesture in South America, there is something moving about seeing the foreman of the A shift share this moment with his son. Like all family rituals in the Avalos home, it's taken on a deeper and richer meaning in the months and years since Florencio was resurrected from the mine.

When he was beginning to die of starvation, Florencio imagined his sons growing up and becoming men and leading the rest of their lives without a father whose cheek they could kiss goodbye. This empty and tragic future has not come to pass.

By the end of my third trip to Chile I've met all the miners save one. Víctor Zamora is not only hard to reach, he also wants a little extra money to talk to me and the men and women producing the movie about the miners. When we finally arrive at his home, just off the highway that leads into the town of Tierra Amarilla, we find a smashed car in the dirt driveway. Zamora opens the door and comes out. The contrast between the confident man who thanked his rescuers in that first video sent up from the mine and this disheveled and disoriented man couldn't be greater. He says he's pawned his wife's jewelry, and the pawn has come due, and he doesn't have the 1.2 million pesos ($2,400) to get the jewelry back. Leopoldo Enriquez, one of the film producers, is also one of Chile's more successful financiers, and he takes a look at the pawn agreement and declares: "This is usury." He agrees to help Zamora pay off the loan, and we enter Zamora's cramped living room.

Víctor Zamora explains that the crashed car outside is his. He's been trying to start a business, buying and selling fruit (there's a load of rotting fruit outside), and this involves a lot of driving back and forth. Recently he was driving on the highway and he blacked out and crashed into a truck. Víctor has been sleep-driving. His subconscious, in an absurdly literal way, is trying to take him back to the mine: He'll start his car and head out for one destination, and slip into a daze, and when he opens his eyes he finds that he's driving on the road to the San José. Víctor explains that his memories of what happened inside that mountain have not stopped their assault on his psyche. "What affected me the most was . . . seeing my own death, and seeing how my companions were dying, slowly," he tells us. At the same time, in the close quarters of the Refuge, he saw the humanity and vulnerability of his fellow workers clearer than he'd ever it seen before. This only made it harder and more painful to watch them approach death. "You see the capacity of human beings to be sensitive in critical moments, how a kind of love is born, a bond [*cariño*], a brotherhood within a moment of danger." Víctor lights a cigarette and smokes steadily as he speaks, and the act of smoking, and of speaking, seems to calm him just enough so that he can tell us what he saw and did in the mine, especially on that first night of hunger.

A few months later, I'm interviewing Luis Urzúa in Copiapó when his phone rings with a call from Zamora. He asks Urzúa if the association of the thirty-three miners, an informal group which Urzúa leads, can loan him a small amount of cash. It's not the first time Zamora has requested such a loan, Urzúa says. When he lived on the streets as a young man, Zamora depended on the generosity and kindess of others, and for some months after his escape from the San José Mine, he returns to that childlike state, reliving his orphanhood—only now with a wife and two children in tow. Eventually, he leaves Tierra Amarilla and returns with his family to the city where he lived on the streets, Arica, some 800 miles and twenty-four hours to the north. He finds work there.

The distance from Copiapó and the mine helps Víctor Zamora. When I speak to him again a year later, he is a man transformed. Long walks on the beach, he says, and listening to friends and relatives speak about their own problems have brought him back to the here and now. I talk to him on the phone, and he sounds like the confident and centered man he was on the first video sent up from the mine. "When someone wants to talk to me, I never say no," he says. He's begun to understand how he can shape his mine memories into something that makes him a better father and husband. "*Queda mucho para vivir*," he tells me. There's still so much left to live.

As deep as Zamora's crisis was, it wasn't as dangerous as the whirlpool of depression and drinking into which Edison Peña has fallen. "One mistake is that we didn't have anything to occupy ourselves with, all that free time wreaked havoc on us," Edison tells me, as he begins to recount the moment he reached a nadir. Since August 5, Edison has been on many journeys, inside the thundering earth and to the cobblestone streets of Jerusalem, among other places, and each time he returns home he drinks. "It's my understanding that the guys who went to the moon afterward just wanted to hang out at a bar, alone. That's the worst thing there is, drinking alone." As the first anniversary of the rescue approaches, Edison's excessive drinking and his statements about wanting to kill himself lead to his confinement in a Santiago clinic. "For my own safety," he says, he was not allowed to leave the well-appointed but small facility, which from the outside resembles the kind of mansion a wealthy Chilean family might call home.

"I'm sitting there for an hour and I want to die . . . I started to feel the horror of being locked in," Edison tells me. He can't help but equate the locked doors of the psychiatric clinic with the stone walls of the mine. "I asked if they would let me out for the September holidays and they said, 'No, the risk is too great.' So I spent my second Eighteenth of

September, the second Independence Day of my country, in confinement." At some point, he was placed in a padded room and in restraints, he tells me, to keep him from hurting himself.

"How long were you restrained?" I ask him.

"I don't remember," he answers. "Don't make me remember. I hate needles and all those things."

His emotional collapse, and the humiliation of the forced treatment that followed, are another challenge to overcome, he says. "After an experience like that, I don't know how it is you're supposed to be in a good mood and show people the positive things about yourself. I think that being able to do that is a gift that comedians have, but I'm not a comedian. What I do think is that Edison Peña is trying to show people something else about himself, something positive . . . The most important thing is being able to talk about these experiences, and to understand that it was something you lived. A lot of people who suffered through an experience like that wouldn't be able to talk about it. If you can talk about it, that's a gift. If you can sit here and talk to a person you don't know very well, and talk about all these things you've been through—that's something. That's courage. It's knowing yourself."

Edison Peña clearly knows himself as well as any of the thirty-three men of the San José Mine, but that doesn't make it any easier to live a healthy life. Months after talking to me he gets drop-dead drunk during a day of meetings with the movie producers at the beach resort of Algarrobo. Several months after that, he travels to Copiapó for a meeting with the other miners convened by Luis Urzúa and the leaders of their association. They see a different Edison Peña, one they've never seen before. "He wasn't drinking," says Urzúa. The man who summoned the will to run inside the mine, to complete a marathon without any preparation, to sing Elvis tunes for strangers in a language he can't speak, is trying to summon the will not to drink. Watching Edison try to live a sober life is, in its way, more impressive than watching him jog in the caverns of the San José Mine.

After months of negotiating, the government grants the oldest miners a retirement pension. It also offers a pension to the younger men, but for them it's not enough to live on, and they turn it down. A few of the younger men get coveted aboveground jobs with the national mining company, Codelco, though these jobs require that they move to southern Chile. Ariel Ticona, whose daughter, Esperanza, was born while he was in the mine, is not prepared to leave the town where he was born and raised. He stays in Copiapó, without a job. Even though he rarely left home, "I didn't want to be in the house, I'd get angry with everybody," he says. Eventually, the man who a year earlier was the most famous new father in Chile leaves his wife and baby girl. "We were separated three, four months. I was always conscious of how I was failing them. I tried to change, but I couldn't," he tells me. After some solitary reflection, "I realized that going back to work was going to be my therapy." Ariel returns to his wife and children, and hears about a local job while playing soccer, from one of his teammates, the miner Carlos Barrios. The job is to operate a vehicle that raises up men using perforators—in an underground mine.

Less than eighteen months after being pulled out of the San José Mine, Ariel Ticona is riding a truck back into an orifice carved into a mountain. "The first day, I felt a little strange," he tells me. "I wasn't scared. I don't know, I just didn't want to be there." Adding to the otherworldliness of the experience is the fact that all his coworkers treat him like a celebrity. "The second day, I got scared. I'd hear machines drilling, and it reminded me of when they were looking for us. By the third day, I started getting used to it." After many sessions in which Ariel has talked to therapists and psychologists about the trauma of being trapped underground, he's forced to enter the dark once more, and after a few hours more in the mine, the undeniable dangers of subterranean work and his ability to conquer his fears (he does not panic and run away) feed the sense that maybe this is really where he belongs. It helps, too, that this is a better and safer mine, one that "isn't too big and isn't too small." "The fourth day, I was starting to like it," he says. Underground mine work has rewards, too, and soon he's earned enough money to buy the house he's been renting, and to start making improvements on it.

Ariel Ticona has come full circle. He is risking his life again to provide a comfortable life to his family. What he has to show for the sixty-nine days he spent in the San José Mine is a top-of-the-line baby carriage from Spain, assorted flags and other mementos, a medal issued by the Chilean Congress, and memories of once-in-a-lifetime visits to Florida, Israel, and a cavernous soccer stadium in Madrid.

For Carlos Pinilla, the onetime general manager of the San José, the legacy of the mine is local ignominy. "I don't look for work in Copiapó," he tells me, in his home in Copiapó. Nor does he ever speak to anyone about what happened in the San José Mine. The day the men were found alive was "the happiest day of my life, happier than the day I was married, happier than the day my first child was born." He knows the miners think of him as the man most responsible for their ordeal, because he's read their postaccident testimony to a congressional commission. When he ran into a miner at an office building in Copiapó they exchanged some tense words: Pinilla says he still doesn't understand why the mine collapsed and never believed it would come crashing down that August 5. But the accident left Pinilla a changed man as he resumed his mining career. He's come to reflect on the person he was. "My treatment of people is friendlier," he tells me, looking beaten down at the end of an hour-long interview. "I don't want to be the ogre-boss anymore. I'm almost begging people to do things with 'please' all the time."

In the months after the rescue, Carlos Pinilla found jobs in other mines, including one that's more than 250 miles south of Copiapó, in Ovalle. As luck would have it, two former workers from the A shift of the San José also found jobs there: Claudio Acuña and José Ojeda. Like Ariel Ticona, they work beneath the surface. Luis Urzúa shares this information with me, and I tell him it seems to me an especially cruel twist of fate: to have one mine fall on top of you, and then to find yourself obliged to work underground in a second mine, with the same boss who once left you behind and trapped for sixty-nine days.

"That's the life of a miner," Urzúa says.

21

UNDER THE STARS

Mario Sepúlveda is riding a horse near his home in Santiago when he's summoned to a meeting with the men and women transforming his story into a book and film. We wait for him in an upper-floor conference room of a Santiago hotel, and when he arrives, late, he's wearing his "kiwi" haircut, jeans, the wool poncho of a Chilean *huazo*, and rubber boots covered in mud. His intense expression and his riding outfit earn him odd looks from the concierge and the bellhops in the lobby. Mario's postrescue life has been busy. A year after his rescue from the San José Mine, the media offers keep coming. He's been invited to star on a Chilean reality television show, one where the participants will spend half their time living in Stone Age conditions, and the other half in the Digital Age. He's started a foundation to build homes for people made homeless by the earthquake and tsunami that preceded the August 5 mining accident. His daughter, Scarlette, has been accepted at the University of Nevada, and he's been invited to speak there, too, and many other places.

In my first interviews with him, Mario wept and told stories of how he and his fellow miners pulled together during the first seventeen days. But now, in these final interviews, we take up the more complicated eight weeks that followed the publication of the letter in which he proclaimed himself "absolute leader," and what happened in the days after he reached the surface, when he gave just enough of an interview to a BBC reporter to allow that reporter to write a book. Many of the miners feel Mario betrayed them and he, in turn, is angry at them, and especially furious with Raúl Bustos.

"I really wanted to kick the shit out of Bustos in the mine, but they never let me," he says.* "But I haven't lost hope of doing it. Here on the surface, one day. I swear . . . I hate the asshole." Mario regales us with mad soliloquies that must be like the soliloquies he unleashed inside the Refuge. He tells us stories and acts them out, standing up to show us how he reacted when he felt the devil's breath on his neck, or falling on the floor to show us what it was like when he fooled Edison Peña with a phony death speech. Above all, Mario raises his voice: in a shout, a plea, a denunciation, a joke. He makes us laugh and he makes us worry for his state of mind. When he talks about Raúl Bustos and his other "enemies" he repeats that same vulgar insult that refers to part of the female anatomy, again and again, despite the fact that his wife, young son, and a female member of the production company are all sitting at the table with us. Many wives would worry to see their husband so angry, but Elvira looks at him with a bemused detachment, perhaps because she knows that Mario can be angry with someone one moment, and love him the next. On those occasions when all the miners are summoned to a group meeting, all his "enemies" become his friends again. Mario can then sit or stand in a room with his brothers from the San José and hug everyone and tell stories as if none of them had ever said an untoward word to one another. In that sense, the sixty-nine days Mario spent in the mine have not changed him, but rather have simply brought the tumultuous aspects of his love-you/hate-you personality to the surface for all the people around him and a worldwide audience to see. He's traveled to California, Germany, Hungary, Mexico, and other places, to be recognized as Super Mario and to offer a few words of his own frenetic brand of Chilean miner optimism. At home, his collection of dogs (strays and puppies) grows to eighteen, he buys a new meat locker (that's always filled), and his wife gives birth to a baby boy, who enters the world after a full term at a healthy eight pounds, eleven ounces. When a Chilean judge finds the mine owners are not criminally culpable in the collapse of the mine, he tells a local reporter: "It makes me

Tenía ganas de sacarle la concha de su madre pero nunca me dejaron.

want to crawl back into a hole underground and not come out." He learns that his hero Mel "Braveheart" Gibson won't play him in the official movie, but that a well-known Spanish actor and heartthrob has agreed to take on the challenging but meaty role of the Man with the Heart of a Dog.

In Copiapó, there is no stardom for Carlos Mamani, the Bolivian immigrant. He's turned down an offer of a good government job in his native country. Instead, like thousands of other immigrants, he decides to try to make a go of it in Chile, his adopted home. He gets a job with a construction company, operating a front loader of a similar make and model to the one he operated for half a day in the San José. One day, he's operating his front loader, dumping dirt into a sifter, when the huge cloud of dust this produces transports him back to August 5 and the San José Mine. "I saw the entire collapse again, just like I lived it those first few moments." He opens the door of his loader and lets out a scream. The sound of his panicked voice wakes him up and brings him back to the present. It's been more than a year since the accident and Carlos is surprised to be having these flashbacks, which cause him to relive the fear and solitude of his weeks underground. "The hardest part about being down there was that I didn't know anybody," he tells me. Today everyone around Carlos knows who he is, but some of his new coworkers are not happy to see a Bolivian man with indigenous features and an Altiplano accent taking one of the company's better jobs. "You shouldn't be here. Why don't you go back to your country," they say. Then they'll insinuate he's wealthy—"a rich Bolivian, imagine that"—and tease him that he should invite the entire crew over to his house for a barbecue. Carlos has dealt with racism before, but never racism tinged with so much envy. The remarks anger him, but his response is to simply keep working. When we meet, his daily routine is taking him to a highway construction project on a road between Copiapó and the next city to the south, Vallenar. He has the satisfaction of getting paid to do the job he began on the day the San José Mine collapsed, only now he's operating a Volvo 150 front loader, a slightly bigger model with "more prestige," he says. When his work on the road is done he'll drive it and take pride in the way he helped link two cities in Chile, the country where he became a Bolivian hero and where he's decided to stay and raise his family.

As with Carlos Mamani, the emotional crisis caused by his sixty-nine days underground hits Víctor Segovia with a delayed effect. More than a year after the rescue, he begins to feel alone and isolated. The unpaid loans he's made have left him feeling used and small, and it bothers him that none of his relatives and friends ask him how he's doing emotionally. It's like they don't care, he thinks. He becomes a recluse, rarely leaving his home, and his inner turmoil begins to manifest itself in physical maladies, including swollen limbs and troubled breathing. His doctors give him medication, but at first it's too strong and it makes him sick. The dose is eventually corrected and this helps Víctor, as does a new outlet he's found for his hurt: He starts writing again, keeping a diary. "My Rescue," he calls it, and in its pages he details the trips he's taken, his descent into depression, illustrating his story and its drama with drawings.

When he was trapped, Juan Illanes kept his mind limber by telling stories and also by imagining himself completing domestic projects. When he gets home to Chillán, he finishes these projects, including that gutter he built several times over in his mind while trying to sleep at Level 105. Among the thirty-three survivors of the San José Mine, he's one of the first to return to work, finding a job with Geotec, the company whose workers and rig drilled the Plan B hole. "If you're working, it's the best therapy for posttraumatic stress," Juan says. Studies have shown that the gravity of posttraumatic stress is directly proportional to the length of time one lives with the threat of death, and Juan slowly unwinds the trauma of the sixty-nine days he lived inside a thundering mountain by going to work, fixing machines, then going back home, and then returning to work again. His job is at the enormous open-pit Collahuasi copper mine, near the port of Iquique, even farther from his home in southern Chile than the San José Mine was, a whopping 1,300 miles in all, or about the distance between New York and Tulsa. Most of his commute is on an airplane, however. He works twelve days on, twelve days off, each dozen days and each journey back and forth across his country causing another layer of hurt to fall away. "I've been learning and growing a lot," he says of his new job.

Months and years pass and the lessons of those sixty-nine days are made clearer to the men of the A shift of the San José Mine. Raúl Bustos remembers how he and his wife were always scrambling for more and better-paying work, and how he was drinking whiskey and Red Bull in the middle of the night, waiting for her to return from another job, when an earthquake and tsunami struck. He remembers the risk he took, for mere money, when he accepted a job in the San José just months later. Now he asks himself: What was I chasing, and why? Sixty-nine days of loneliness liberated him from his restlessness. His family, his home, the garden in the back, and the peach tree—he sees them all bathed in a brighter light, and time moves slower and seems richer somehow. "We're not so worried about getting ahead," he says. Pedro Cortez, the young miner who didn't like the "rock star" treatment he got in the months after the rescue, starts to feel better when the pressures and expectations of his celebrity start to fade. The nightmares and daydreams that caused him to weep go away. "You start thinking about other things," he says: his daughter, his studies, the new jobs he might get, the new person he will become. "You don't think about the tragedy anymore."

Darío Segovia and his wife, Jessica, launch their own business. Instead of buying and selling produce as Darío had once hoped, they become distributors for a soft-drink company. Darío maintains their two-truck fleet and Jessica does the accounting and their home is their office, and they meet their employees there for breakfast every morning. Darío and Jessica go to bed late and get up early. "Movie money or no movie money, we're going to keep this business going," Jessica says. She likes the idea that they've become "micro-entrepreneurs," following in the footsteps of Darío's sister María.

"There's nothing like making your own dough," María says. "If what you have you earned in sacrifice, you value it more." The onetime mayor of Camp Esperanza keeps working in Antofagasta, and she's selling her pastries at a flea market one day when one of her daughters calls and tells her: "Mommy, you're going to have to be strong again." Her thirty-six-year-old daughter, Ximena, has been diagnosed with leukemia. The disease advances quickly, requiring Ximena's transfer to an intensive-care unit in Santiago, and once again María Segovia feels as if the forces of nature and fate are trying to sweep her family away. She takes to the bus and travels south. María's daughter is dying, and like any mother she will do anything to help her, so when she gets to Santiago she calls the most powerful person she knows, Minister Golborne. The minister receives her in his office, and calls her his friend, and embraces her, and listens to her story, and weeps with her. He promises to call the minister of health to make sure Ximena is getting the best possible care, and not only does Golborne make good on his promise, he shows up at Ximena's hospital with flowers and with the minister of health, too. Ximena gets better.

A cynic might note that Minister Golborne is about to launch a presidential campaign and that he visits Ximena in the hospital with the Santiago media in tow—the coverage is a reminder of Golborne's role in the miracle of the San José Mine. Golborne's compassion and tenacity during the search and rescue at the mine made him one of Chile's most popular politicians. The Piñera administration has disappointed most Chileans, but Golborne is a rising star, a conservative leader with a compelling story of service to thirty-three workingmen and their families. Two years after the rescue he seems poised to easily claim his party's presidential nomination. But his campaign is quickly brought down by controversies, including one that completely undermines his image as a man of the people: the revelation that he kept part of his wealth in an undeclared account in the British Virgin Islands.

Events in Santiago always seem very far away to Juan Carlos Aguilar, the former supervisor of the mechanics crew. He's returned home to the southern edge of mainland Chile, in the town of Los Lagos, and leads a low-key life that includes giving talks to local schools. In his gentle and unhurried voice, and his unassuming demeanor, he tells children and teenagers about the importance of teamwork, about how humble people can overcome the steepest odds, and how faith can give you strength even when death is near. His role as one of the leaders of the thirty-three men underground is not widely known, but his fellow miners all acknowledge what he did, and that's enough for Juan Carlos. Two or three times a year he travels half the length of Chile to meetings in Copiapó, because those same men have chosen him to be one of three members of their leadership committee. They discuss starting a foundation to help impoverished miners. But Juan Carlos feels he isn't doing enough. Something special happened to him, something that nearly killed him and that gave him new life. He remembers the faces of the men in that hole, and the slate-colored "guillotine" that trapped them. In letters written in the Refuge and in starvation dreams, they said farewell to their families, and then an entire country worked to bring them out of the dark. Juan Carlos wonders why he, among so many millions of workingmen, was chosen to live these things. "I wake up every morning and ask God, 'What am I supposed to do with this?'"

Two and a half years after the accident, Juan Carlos Aguilar is alongside Luis Urzúa in Copiapó at the press conference announcing the creation of the miners' new nonprofit foundation, the Thirty-Three of Atacama. The foundation's mission is to aid the poor of the region, and the *perquineros*, the artisan miners whose culture shaped the childhood and family life of several of the thirty-three men. Most of the survivors of the San José Mine are there, including Mario Sepúlveda, Raúl Bustos, and Edison Peña, along with the former minister of mining Laurence Golborne, who can no longer be accused of having immediate political motives. They meet at the Antay Casino and Hotel, across the street from the chapel that holds the Catholic relic most sacred to local miners, and they listen as Luis Urzúa reads some prepared remarks. "Why did we survive a tragedy?" he asks. "What is the mission each one of us must follow?" Since the rescue, Luis has taken classes in public speaking, and aboveground he's tried to be something he wasn't underground—the undisputed leader of the thirty-three miners. He tells me he's learned to talk to "big people," and more than once he's written pointed letters and e-mails to powerful lawyers and film producers in Santiago and Los Angeles. Eventually, however, his fellow miners question his actions (much as they did underground) and they vote to remove him from their leadership committee.

The last time I speak with Luis, he is angry with many of his former colleagues. But to outsiders he still retains his role as a figure of Chilean and mining history. Luis can feel the way the collapse of the mine and the drama of the rescue linked him to strangers around the world. In Chile, people tell him they'll never forget where they were when they heard the news that the men had been found alive. It was on a Sunday, a family day, at the hour of the day's biggest meal. "We were just sitting down to eat," they tell him, and they remember the church bells ringing, and the people running out into the streets. Chileans who were traveling abroad when 1.2 billion people watched the men being rescued on television tell him that to be Chilean in those days in October was to feel the entire world had adopted you. People congratulated us for being *chilenos*, they tell him.

From all these strangers, Luis does not get the sense that they think he's a hero, necessarily, or that they're in awe of him. Rather, he understands that it's as if he and these strangers had lived something together: a shared experience with him in the mine and them on the outside. What he feels from these strangers is the gratitude of people who've been given a true and hopeful story, a timeless legend born of their own time, in a humble country in the shadow of the Andes.

The first time I met Alex Vega, it was in his home in Copiapó, about eleven months after his rescue. He was not well. I asked him about his family history, and how it was he came to work in the San José, and I noticed his hand was trembling. I talked to him a bit more, and the tremor seemed to spread through his body, as if he were shivering from cold on that warm, sunny day of Southern Hemisphere spring. "You look a little nervous," I said, and soon Alex talked about the "scars" he still had and how he'd struggled with mood swings, and nightmares that were starting to diminish in frequency but which still haunted him. Above all, what troubled Alex was the sensation that he needed to be alone, that his wife and children were better off when he wasn't in the same room with them, which is a painful and confusing thought for a family-loving man to have in his head, especially after spending sixty-nine days longing to be reunited with them.

Eventually, Alex's response to his emotional crisis is to hold on tighter to his wife, Jessica, and to depend on her more. He realizes he can't leave home for a long trip without her, and when he goes to Santiago for a meeting with the other miners, it's the thirty-three men and Jessica Vega, not because Jessica is a busybody but simply because Alex isn't able to get on a plane or be in a hotel room without her. The other miners don't hold this against him, because they all know exactly what he's going through.

Each time I meet Alex again—seven months later, a year later—he's doing better. At his home in the Arturo Prat neighborhood, I meet his father and his sister and brother, and I interview Jessica, too. After a while, it's clear to me that there's a lot about what happened to Alex underground that Jessica doesn't know. "Can I tell her?" I ask. Eventually, she'll read it in this book, and maybe it's better she hear it from me now. Alex says it's okay, so I tell Jessica how Alex was sent flying by one of the blasts caused by the mine's collapse; I tell her how, in his desperation to reach home, he risked his life trying to crawl under the stone blocking the Ramp; and I tell her how he offered not to eat on the sixteenth day, when the other miners said he looked painfully thin, weak, and hungry. The realization that her husband has been carrying these memories as he walks about their home causes Jessica to weep, briefly, and then later, perhaps, it brings her a measure of understanding, as if the plot and meaning of the movie in which she'd been living had suddenly been made clearer to her.

The accident at the San José Mine briefly made Copiapó the most famous mining city in the world. In the years since, life there has quickly returned to its everyday rhythms, to the routines of mineral extraction and the vagaries of weather and geology Darwin noted in his journal. An 8.2 magnitude earthquake shook the Atacama Desert. And nine months after the thirty-three men of the San José Mine were rescued, torrential rains fell on the city. The Copiapó River filled with water for the first time in fourteen years, and the riverbank flooded, forcing the evacuation of the Tornini neighborhood of squatters, much as another storm had driven María and Darío Segovia from their home as children. The Tornini squatters were later evicted by the city government to make way for an access road to a new nearby riverside mall that is yet another symbol of the city's booming growth. After the neighborhood had been emptied but not yet demolished, stray dogs wandered through the ruins.

About a hundred yards away, on the other side of the bridge that takes the Pan-American Highway across the Copiapó River, the city erected the most conspicuous local monument to the rescue of "the thirty-three of Atacama," a tall, chrome-skinned woman holding a dove. Donated by the Chinese government, the sculpture faces the dry bed of the Copiapó River and greets all those who enter Copiapó in cars and buses from the south, including those who come to work in the region's mines. Outside the city, at the San José Mine, a cross and monument have been erected at the spot where the families of the thirty-three men gathered and built Camp Esperanza. Only rarely do tourists undertake the forty-minute drive from Copiapó to visit. The entrance to the mine is covered, top to bottom, with a chain-link fence. If the guard at the site isn't looking, you can walk up to the fence and peer into the gloomy tunnel that leads into an empty and broken maze of caverns below.

The last time I meet with Alex Vega and his family, it's not to interview them but to share a meal together. I head out on foot to their home and enter a working-class neighborhood of Copiapó at twilight, a collection of homes hugging the low ground. There is no one else but me on the sidewalk, until I see a group of young men gathered under a weak streetlamp that's just come on in the growing darkness. They look furtively down an asphalt avenue conspicuously absent of traffic, and I walk past them, and then past warehouse buildings, and homes of tin and wood clustered behind tall concrete walls. Once again I am alone on the street, though I sense that behind these barriers there are especially industrious families who have filled their properties to the brim with rooms, furniture, and appliances—but whose humility allows them to enjoy this prosperity only if they can be certain no one will see them. I take a wrong turn and come upon a pair of boys rolling a tire down one of the sloped streets; they help me find my way.

When I reach Alex's address I find his homestead is how I remember it: unfinished. Alex hasn't completed the building project that sent him into the San José Mine to earn a bit of extra money. He's still got one old room that's crowded with stoves and a table, and one newer room with a big couch and some sofa chairs; in between, there's an empty, open space waiting to be filled with new construction. The wall he's been building with his wife is taller, however, and nearly done. When he shakes my hand I note how, for the first time, it's with the firm grip of a man who works with tools for a living. After more than two years of emotional suffering, Alex has taken several steps to heal himself, including returning to work, at a job repairing vehicles. When we sit and talk, he tells me what he did to end his nightmares about being buried alive. "I couldn't sleep, so I told myself, 'I need to confront this fear. I need to go back into a mine.'" He asked a brother-in-law who worked underground if he could join him, and for a week Alex entered the mine every day, going three hundred meters below the surface. He drove down into the deep dark, wandering about the stone passageways, and then back up and out those dank caverns and into the sunshine. The nightmares never returned after that, and he stopped waking up and crying in the middle of the night. The next step in his recovery, he tells me, will be to host a gathering of all his family and friends, to talk about what he saw and survived during his sixty-nine days underground, a topic that he and the people around him have deliberately avoided for years. "I want to turn the page and leave it all behind."

When Alex's sister Priscilla and her boyfriend, the mariachi Roberto Ramirez, arrive, and Alex's brother Jonathan arrives a bit later, the mood is light and full of laughter. The presence of the author writing the book about Alex and his coworkers leads Jessica, Roberto, and Priscilla to remember what Camp Esperanza was like, with its bonfires and its families and its odd characters: the clown who came to make the children laugh, the celebrities who came to have their pictures taken, the workers who came to drill and search, and the reporters who came from every corner of the globe. Then, because it's a pleasant night, and because some of the Vegas want to smoke, we go outside.

As his two small children run and circle him several times in the chasing game they're playing, scampering in and out of the cement courtyard, Alex has the calm, content look of a man who has returned home from a long journey, and who can see how his presence is soothing and a source of strength to the people around him. Under the black sky and the stars, I listen as his family tells more stories about the sixty-nine days they spent at the San José, and especially about the predawn hours of August 22 in Camp Esperanza. Remember that night, Roberto asks, what a magical night that was? Alex says no, he doesn't remember, and everyone laughs: Well, it *felt* like you were there with us, Alex, even though you were still buried seven hundred meters underground. It was a cold August night, but the Vega family was full of hope, because the Plan B drill was said to be close to the Refuge, and they had faith it would break through to Alex. The desert around the mine was covered with flowers, after a rare shower a few days earlier. The Vegas remember the songs they sang that night, including the one that Roberto wrote about "El Pato" Alex and his seventy-year-old father entering the mountain to search for him.

On that night, in a flower-covered desert, and in a fungi-filled cavern underneath, Alex and his family lived an epic story that belongs to the world and to the history of Chile; but it's also a family tale as intimate as this small and still-unfinished space Alex and Jessica Vega call their own. For Alex, the odyssey ends with the renewal of the daily rhythms in his home: on mornings when he leaves for work as the fog called the *camanchaca* rolls in; on afternoons when the desert sun burns through it; and on nights like this one when the cooling air calls him outside with his family. I look up at the unfamiliar canopy of the Southern Hemisphere night above me, and I see a sky similar to the one Alex's family saw from Camp Esperanza, including a constellation called Phoenix, and the five bright stars of Crux, a cluster of stars also known as the Southern Cross. Under the southern stars, before dawn on August 22, they had faith they would soon witness a miracle, and they sang a song that declared Alex would be freed from his mountain prison. Tonight they sing it again, for a visitor from a faraway country, and for Alex.

"And El Pato will return!"

"He will return!"

When the song finishes, Alex Vega looks at the people who love him, and who are smiling with the memory of the night they first sang those words.

"And here I am," he says.

ACKNOWLEDGMENTS

This book is based, in large measure, on the hundreds of hours of interviews I conducted in Chile with the thirty-three survivors of the San José Mine collapse and their families, and on the diary kept by Víctor Segovia. Many of the interviews were conducted in the homes of the miners, individually and privately, or with their wives, girlfriends, children, and other relatives present. Some interviews were conducted in groups, and in hotels in Copiapó, and in the offices of the Carey law firm in Santiago.

As with any life-threatening and life-altering experience lived by a large group of people, there are often dramatic differences of opinion about certain events, and about how certain individuals conducted themselves. The thirty-three men, as a group, entrusted me to tell their story fully, and to sort out exactly what is truth and what is myth, and I've tried to do so to the best of my ability. The responsibility for any errors in the text is mine alone.

Many of the physical details of the miners' experiences during their first seventeen days trapped underground are derived from my interviews of the miners—but also from several videos shot by the men themselves. José Henríquez took his cell phone into the mine that fateful August 5 (instead of leaving it inside his locker on the surface, as others did), and that cell phone's camera provided the only visual record of those days. In addition, the Chilean government shared with me the unedited version of the first video shot not long after the men were discovered, footage that provided dramatic evidence of the physical degradation of the men and of the inhuman conditions inside the San José Mine. I also had access to additional private images and videos shot by the miners themselves (with cameras sent down after rescue shafts reached them). In reconstructing what it's like to enter and work in a subterranean mine, I also benefited greatly from a visit to the interior of a nearby mine with the then minister of mining, Laurence Golborne.

In addition to my interviews, a key document in writing this book was the report on the accident and its causes prepared by an investigative commission of Chile's Congress. The NASA experts who traveled to Chile told their stories to a NASA oral history project, and I drew upon those accounts in the chapters on the rescue effort. More details came from the interviews the American driller Jeff Hart granted to a Colorado television station and a talk he gave at the Colorado School of Mines. A 1993 study of the geology and "mineralization" of the southern Atacama by several scientists for the *International Journal of Earth Sciences* (*Geologische Rundschau*) was a source for the passages on the geological history of the region. Darwin's reminiscences of Copiapó and the Atacama are from his famous journal *The Voyage of the Beagle*. The account of Pedro Rivero's unsuccessful rescue attempt is drawn, in part, from his November 2010 interview with the mining magazine *Area Minera*. The excellent 2011 retrospective on the rescue by the industry magazine *Minería Chilena* also provided and confirmed critical facts. Many details about the rescue come from the voluminous daily coverage of the events in Chilean newspapers, especially from *La Tercera* and *El Mercurio,* and from the work of Carlos Vergara Ehrenberg in his book, *Operación San Lorenzo.* This longtime veteran of newspaper reporting would like to acknowledge the professionalism of the Chilean writers and photographers who covered the San José Mine disaster, and he hopes that they will see the influence of their collective labors on this account.

Among the many interviews I conducted with rescuers and officials, many stand out, especially those with Golborne, Cristián Barra, Pedro Gallo, André Sougarret, the drillers Eduardo Hurtado and Nelson Flores, the rescuer Manuel González, the shift supervisor Pablo Ramirez, and Carlos Pinilla. The psychologist Alberto Iturra said he spoke to me because "my clients, the miners," asked him to do so. And, finally, I am especially indebted to the families of Darío Segovia, Florencio Avalos, and Alex Vega for the many hours they spent with me in their homes.

The attorney María Teresa Hola guided me around Copiapó when I was just learning to navigate the city, and also shared her knowledge of her native city's history, as did many other Copiapó residents. At the Carey law firm in Santiago, Paulina Silva and Pilar Fernández recounted their experiences with the miners to me, and Claudia Becerra and Soledad Azérreca helped track down the miners for interviews and organized the logistics of my trips to Chile. The attorneys Guillermo Carey, Fernando García, Remberto Valdés, and Ricardo Fischer were among those who agreed to entrust me with this project, and I am grateful for their support.

At Phoenix Pictures, Edward McGurn, Patricia Riggen, and the legendary Mike Medavoy offered many wonderful words of encouragement after reading my manuscript when it was still a work in progress. Nuria Anson transcribed hundreds of pages of my interviews—without her tireless and exceedingly fast work, this book would have taken years longer to write. Jessica Boianover in Buenos Aires transcribed additional interviews, as did Jazmin Ortega in Los Angeles, and Ricardo Luis Mosso in Buenos Aires conducted research. Idra Novey provided much helpful advice on my Spanish translations. The screenwriter José Rivera was my partner during many interviews, and his insights on the miners' story were invaluable. The film producers Leopoldo Enríquez and Cecilia Avalos also helped set up many interviews.

I wrote most of this book while employed at the *Los Angeles Times*, and I'd like to thank the colleagues who helped me fulfill my responsibilities there while I simultaneously took on this massive project, including Nita Lelyveld, Joy Press, Carolyn Kellogg, and David Ulin. And I am deeply indebted to Judy Baldwin for her insights about the creative process and the human soul: Her counsel helped keep me sane and centered while writing this book.

Thank you to Jay Mandel, Alicia Gordon, and Eric Rovner at William Morris Endeavor for bringing this project to me and entrusting me with it. My longtime friend and editor Sean McDonald didn't flinch when I told him I wanted to write this book, and he gave it a home at FSG.

Finally, and most important, my wife, Virginia Espino, sustained our family for the three years I spent working on this book, and endured my absences during the five trips I made to Chile. I could not have written this book without her love and support. Thank you, *amor de mi vida*.

Join a literary community of
like-minded readers who seek out
the best in contemporary writing.

From the thousands of submissions Sceptre
receives each year, our editors select the books
we consider to be outstanding.

We look for distinctive voices, thought-provoking
themes, original ideas, absorbing narratives and
writing of prize-winning quality.

If you want to be the first to hear about our
new discoveries, and would like the chance to
receive advance reading copies of our books
before they are published, visit

www.sceptrebooks.co.uk

Follow @sceptrebooks

'Like' SceptreBooks

Watch SceptreBooks